Pour ma

et courageuse
Tante Colette

avec beaucoup de
tendresse

Winny

Sustainable Consumption, Ecology and Fair Trade

Sustainable consumption is an objective attracting growing attention within sustainable development policy circles, academic research and society more generally. This timely volume discusses the key debates concerning the involvement of consumers in this field.

How can we analyse the roles of consumers, their motivations and effectiveness relating to sustainable consumption? What does sustainable consumption represent for consumers, in particular from the standpoint of ecology and fair trade, and which are the changes taking shape that can give form to more effective policies in these areas? These are some of the important questions addressed in this work. The contributors examine a range of interesting and relevant case studies including: household energy consumption; the promotion of green products; Fair Trade; Oxfam Worldshops and Southern producers.

This book takes an interdisciplinary approach, including sociology, policy studies, philosophy, engineering and marketing. This well-balanced study presents theoretical debates, as well as empirical evidence in order to:

- Characterize the basic problems and determiners of an evolution towards, and the obstacles to, more sustainable consumption patterns.
- Produce knowledge on the profile and motivations of consumers sensitive to these issues.
- Explore modes of interaction and innovation for changes in which consumers are involved.

This text will be of key interest to those researching and studying in the fields of sustainable development, Environmental Politics and Sociology.

Edwin Zaccaï is Professor of Socio-politics of the Environment and Director of the Centre of Studies on Sustainable Development at the Free University of Brussels, Belgium.

Environmental Politics / Routledge Research in Environmental Politics

Edited by Matthew Paterson, *University of Ottawa* and Graham Smith, *University of Southampton*

Over recent years environmental politics has moved from a peripheral interest to a central concern within the discipline of politics. This series aims to reinforce this trend through the publication of books that investigate the nature of contemporary environmental politics and show the centrality of environmental politics to the study of politics per se. The series understands politics in a broad sense and books will focus on mainstream issues such as the policy process and new social movements as well as emerging areas such as cultural politics and political economy. Books in the series will analyse contemporary political practices with regards to the environment and/or explore possible future directions for the 'greening' of contemporary politics. The series will be of interest not only to academics and students working in the environmental field, but will also demand to be read within the broader discipline.

The series consists of two strands:

Environmental Politics addresses the needs of students and teachers, and the titles will be published in paperback and hardback. Titles include:

Global Warming and Global Politics
Matthew Paterson

Politics and the Environment
James Connelly & Graham Smith

International Relations Theory and Ecological Thought
Towards Synthesis
Eric Laferrière & Peter Stoett

Planning Sustainability
Edited by Michael Kenny & James Meadowcroft

Deliberative Democracy and the Environment
Graham Smith

EU Enlargement and the Environment
Institutional change and environmental policy in Central and Eastern Europe
Edited by JoAnn Carmin and Stacy D. VanDeveer

Routledge Research in Environmental Politics presents innovative new research intended for high-level specialist readership. These titles are published in hardback only and include:

1. The Emergence of Ecological Modernisation
Integrating the Environment and the Economy?
Stephen C Young

2. Ideas and Actions in the Green Movement
Brian Doherty

Sustainable Consumption, Ecology and Fair Trade

Edited by Edwin Zaccaï

Routledge
Taylor & Francis Group

LONDON AND NEW YORK

First published 2007
by Routledge
2 Park Square, Milton Park, Abingdon, Oxon OX14 4RN

Simultaneously published in the USA and Canada
by Routledge
270 Madison Ave, New York, NY 10016

Routledge is an imprint of the Taylor & Francis Group, an informa business

© 2007 Edwin Zaccaï for selection and editorial matter;
individual contributors, their contributions

Typeset in Garamond
by Keystroke, 28 High Street, Tettenhall, Wolverhampton
Printed and bound in Great Britain
by Biddles Digital, King's Lynn

British Library Cataloguing in Publication Data
A catalogue record for this book is available from the British Library

Library of Congress Cataloging in Publication Data
Sustainable consumption, ecology and fair trade / edited by Edwin Zaccaï.
    p. cm.
    Includes bibliographical references.
    1. Consumption (Economics)–Environmental aspects. 2. Consumption
(Economics)–Moral and ethical aspects. I. Zaccaï, Edwin.
    HB820.S85 2007
    339.4'701–dc22   2006021849

ISBN10 0–415–41492–X
ISBN13 978–0–415–41492–0

In New York and Paris, on the Champs Elysées
They see her coming from a long way
They clap their hands together when they get her in their store
She's gonna wanna get more more more and more and more
Imelda baby, Imelda baby what to do
All the poor people saying that they gotta quit paying for you

(Mark Knopfler, Imelda)

Contents

Illustrations

Boxes

Figures

Tables

Contributors

Françoise Bartiaux is a Senior Researcher in Sociology for the The National Fund for Scientific Research and associate professor at the University of Louvain at Louvain-la-Neuve, Belgium.

Paul-Marie Boulanger is Director of the Institute for a Sustainable Development, Ottignies, Belgium.

Patrick De Pelsmacker is Professor of Marketing at the University of Antwerp Management School, Belgium. **Wim Janssens, Caroline Mielants** and **Ellen Sterckx** are researchers in this School.

Michelle Dobré is a Researcher at the Centre Maurice Halbwachs (CMH) and Lecturer in Sociology at the University of Caen, France.

Nadine Fraselle is a Senior Researcher in Sociology at the Centre Entreprise-Environnement at the University of Louvain at Louvain-la-Neuve, Belgium.

Serge Latouche is Professor Emeritus in Economics, University of Paris-Sud (XI).

Ronan Le Velly is Lecturer in Sociology at the University of Nantes, France.

John Lintott is a former Senior Lecturer in Economics at South Bank University, London.

Gautier Pirotte is Lecturer in Social Anthropology at the University of Liège, Belgium.

Catherine Rousseau heads the division of Consumption and the Environment, in the Centre for Research and Information of Consumers Organizations, Belgium. **Christian Bontinckx** is a Consultant in Psychology.

Coline Ruwet is a Ph.D. student in Sociology at the Hoover Chair of Economic and Social Ethics, Catholic University of Louvain at Louvain-la-Neuve, Belgium.

Isabelle Scherer-Haynes is a Researcher in Social Sciences, University of Liège, Belgium.

Anton J. M. Schoot Uiterkamp is Professor of Environmental Sciences at the Centre for Energy and Environmental Studies, University of Groningen, The Netherlands.

Grégoire Wallenborn is a Researcher in Science, Technology and Society Studies at the Centre of Studies on Sustainable Development, Free University of Brussels.

Edwin Zaccaï teaches the Socio-politics of the Environment and heads the Centre of Studies on Sustainable Development at the Free University of Brussels and is a Lecturer at Sciences-Pô in Paris.

1 Introduction

Contradictions and studies

Edwin Zaccaï

Policies for sustainable consumption

A number of public policy areas related to sustainable development have in recent years brought changes in consumption and production patterns to the fore. This expression was highlighted in Chapter 4 of Agenda 21 (UNCED, 1992). This chapter was brief in part because it gave rise to conflict, particularly the idea that the consumption patterns of the developed nations would not be sustainable: 'In many instances, this *sustainable development* will require reorientation of existing production and consumption patterns that have developed in industrial societies and are in turn emulated in much of the world' (4.15).

That key idea was translated ten years later into the United Nations programme adopted in Johannesburg (2002) which in these matters: 'Encourage and promote the development of a 10-year framework of programmes . . . to accelerate the shift towards sustainable consumption and production' (§14).[1] Meanwhile, this objective has gained importance and is associated with policies along these lines in bodies such as the Organization for Economic Cooperation and Development (OECD) (OECD, 2002a) and the European Environment Agency (EEA) (EEA, 2005).

In these contexts, the many initiatives, declarations and watchwords launched in recent years by public policy makers, companies or activist movements interested in these questions expect *consumers* to exert a significant favourable influence. Reasons for this evolution are linked to the proportionally growing ecological impact of consumers' habits in certain fields; the framework of the growing merchandization of the economy; the move from the public powers as central regulator to actors within schemes of governance; the individualism of 'consumer-citizens'; not forgetting the tradition of activism through example.

How can we analyse these roles, their motivations and effectiveness? More broadly, what does sustainable consumption represent for consumers, in particular from the standpoint of ecology and Fair Trade, and what changes are taking shape that can give form to more effective policies and actions in these areas?

These are the key questions addressed in this book, which is the result of a dialogue and pooling of several research projects aiming to:

- characterize the basic problems and determiners of an evolution towards and the obstacles to more sustainable consumption patterns;

- produce knowledge on the profile of consumers sensitive to and not sensitive to these issues;
- explore realistic ways of interaction and innovation for changes in which consumers are involved, in interaction with consumption conditions, the actions of producers and initiatives by the public authorities.

A majority of these research projects were carried out in the context of providing scientific documentation as possible support for public policies related to sustainable production and consumption patterns.[2] That does not mean that definitions resulting from these political projects are taken for granted in this book, and the objectives themselves will be analysed critically. Since 2000, a number of academic books and synthesis reports (Heap and Kent, 2000; Cohen and Murphy, 2001; Princen *et al.*, 2002; Jackson and Michaelis, 2003; Reisch and Ropke, 2004; Southerton *et al.*, 2004; Jackson, 2006) have offered rich insights for the exploration of key questions in this respect, focusing especially on ecologically sustainable consumption. In this book, the issue of Fair Trade has been combined with the former to consider sustainability in broader terms.

These questions are more or less present in most of the industrialized countries and, if much of the field research has been accomplished in Belgium, the angle of our work is decidedly European in the ways it is put into perspective and, thanks to the contributions of academics from different countries, comes from a number of disciplines. I will return below to the way their contributions tie in with one another. However, I would first like to highlight contrasts that may make clear straight away the considerable difficulties that are masked by what at first sight appears to be a consensual position.

A short list of contradictions

By way of introduction, I will thus try to formulate the questions at stake in the form of a list of contradictions. My hypothesis here is that the question of sustainable consumption stands at an essentially *contradictory stage*. These contradictions act as powerful brakes on changes aimed at achieving the stated objectives. For some of these contradictions, it is hard to see how they might be resolved in keeping with the objectives. Highlighting them may consequently serve as a reminder of the backdrop against which actions are being initiated. For others, there are partial options where changes can take place, and the different papers try to clarify a certain number of them. In any case, our academic privilege is perhaps to assume the possibility of pointing out the existence of conflicts and contradictions, without claiming that, for sustainable consumption as it stands today, there are already policies capable of surmounting them. In this respect I adopt a critical approach more than a problem-solving approach. For each contradiction I nevertheless make a few suggestions, concluding, with the last, with a summary assessment of the general state of this problem today. These introductory analyses are grouped first by three

---

**Box 1.1   Ten contradictions on the road to
sustainable consumption**

1   Economic growth constitutes a basic foundation of economies, which
thwarts limits to consumption.
2   Growth in consumption is still identified with a model of well-being.
3   There is an accelerated adoption of high consumption standards in
rapidly industrializing countries.
4   Consumers take advantage of high levels of competition to put
downward pressure on prices, which discourages more costly – and
sustainable – production standards (both social and ecological).
5   Discriminating between products, to identify those that meet
sustainable consumption requirements, demands means of analysis
lacking to consumers.
6   Attitudes in support of sustainable consumption are translated into
only limited reductions of the negative impacts of consumption.
7   From the standpoint of enterprises, at ecological level the green
profile of products is not a first order positioning factor, while at
social level voluntary commitments reflect limited changes.
8   Information instruments are preferred tools for change even though
their impact is weak.
9   The objectives of a sustainable consumption policy are relatively
vague.
10 Although the goal of sustainable consumption raises many questions,
the issue appears consensual rather than being the subject of policy
debates.

---

relating to the basic determiners, or macro-drivers (points 1–3), after which
I look at the role of consumers (points 4–7), followed by the role of public
policies (points 8–10).

Economic growth constitutes a basic foundation of economies, which thwarts limits to consumption

We might begin by pointing out the somewhat incongruous nature of motivations for limiting consumption in societies where, on the contrary, consumers
are encouraged to support economic growth, which is needed to reward investors
and permit the sharing of certain economic benefits in society. Initiatives
such as the 'no-shopping day' are tolerated at present because they are relatively
limited in scope, but how could they expand without at least prompting
impassioned debates? Some voices are advocating for the end of growth in
rich societies, but should it come from documenting its numerous deadlocks,
as in the substantial book by R.U. Ayres (1998), or in a movement towards

'de-growth' (see Latouche, Chapter 12, this volume), though this catchword taken as a project (a radical 'update' of the sustainable development motto) appears at the moment to be rather ambiguous and ill-defined.

In contrast, what is very clear is that from a macroeconomic standpoint, businesses and political leaders work actively to reduce the costs of production and of commercial transactions, the result being a sought-after growth of production and trade. Looking at the ecological consequences of this deep economic driver, a 'rebound effect' may be observed in many cases, such as transport. This effect means that even if individually more 'eco-efficient' products are consumed, their growing quantity implies that their cumulative impact in general does not decrease (Røpke, 1999), not to mention the much more common cases where there is no specific effort to reduce the ecological impacts of the products themselves.

In this context, ecological objectives are brought down to a 'second best', namely the decoupling of economic growth from environmental impacts (OECD, 2002b). But this decoupling is only taking place in certain zones and for certain vectors. In addition, from an ecological point of view it can only be a secondary objective: it is after all the absolute value of impacts that should diminish, not their relative value compared to the (growing) economic output. According to ambitious studies (Sachs *et al.*, 1998; von Weiszäcker *et al.*, 1997) and macroscopic environmental research (Vitousek and Mooney, 1997), as in international political programmes,[3] reductions in some of the impacts should nevertheless be highly significant in the long term to ensure the ecological sustainability of the planet. Difficulty complying with the Kyoto objectives, a major aim of decoupling, provides a symbolic illustration of the extent of this challenge.

Growth in consumption is still identified with a model of well-being

The previous contradiction looked at material economic drivers. This one is about the cultural associations between growth in consumption and in happiness. Consumerism has been identified since the beginnings of ecology in the late 1960s as the central obstacle to limiting the impacts of contemporary societies (Gorz, 1977; Bozonnet, 2005). In fact, this criticism dates back much further: 'Consumption is not geared unfailingly towards increasing our well-being and improving our quality of life – as economics would have it. Some of it appears to impede well-being and decrease our quality of life. This kind of claim has occupied a central role in critiques of society for well over two centuries', write Jackson and Michaelis (2003: 27) in a chapter entitled 'Consumerism as a social and psychological pathology' (see also Wilk (2001, 2002) and Dobré (Chapter 11, this volume). John Lintott (Chapter 4, this volume) updates this argument by means of satisfaction surveys that largely confirm the claim. So how can we act on the values that underpin these behaviours (Brown and Cameron, 2000)?

The effort to 'change the indicators' (e.g. to stop giving the central role to economic growth as a measure of progress), represents a constant in recent years in the field of sustainable development (Max-Neef, 1995), or more broadly in social economics (Viveret, 2002). While it is certainly necessary, it requires considerable means and numerous developments to impact concretely on the differentiated behaviours of actors beyond a general measure of well-being such as that of the UNDP Human Index Development, for example.

Another non-exclusive possibility consists of making a better distinction between segments of the population in whom these objectives strike a chord, knowing that the dissociation between increased consumption and well-being covers countless realities. However, such dissociation can only be partial.

There is an accelerated adoption of high consumption standards in rapidly industrializing countries

This third difficulty underlines the contemporary context of consumption patterns at the global level. Analysis shows that the fear of ecological impacts associated with considerable projected consumption increases at global level constitutes a powerful historic root of sustainable development conceptualization (Zaccaï, 2002). Inversely, as W. Sachs puts it, 'sustainability implies creating a style of wealth which is capable of justice' (Sachs, 1999a). Global studies (Millennium Ecosystem Assessment, 2005) show that this direction has not been taken in the past fifty years. Discourse on revising growth in consumption as a model for lifestyle therefore appears out of sync, in a period where these models are being rapidly replicated. Calls for limits directed towards the fringes of the population having access to these new standards may be interpreted as eco-colonialism, and cannot be taken seriously given the massive historic 'addiction' to the same consumption patterns in economically rich countries. A survey of data published about the 'new consumers' in newly industrialized countries (Myers and Kent, 2004) confirms and reactivates this fast-evolving problematic. And yet, at the very least, sustainable consumption must be analysed in the context of globalization (Fuchs and Lorek, 2001).

At an ecological level, important questions hang in the balance and will soon be played out in the choices and standards adopted by the rapidly industrializing nations, China first and foremost. Not only will this determine their own impacts, but as the 'planet's workshop', the standards could influence world development more widely (Izraelewicz, 2005). For certain impacts, the serious level of pollution already reached may sometimes act 'favourably' in terms of promoting more advanced standards.[4]

Questions also arise with regard to the social aspects of this phase of industrialization, to which we will return in some of the following points.

Consumers take advantage of high levels of competition to put downward pressure on prices, which discourages more costly – and sustainable – production standards (social and ecological)

In short, this fourth contradiction lies between the individual's interests as a consumer and as a worker. This contradiction is not new and underpins the law of supply and demand, but we may observe a change of context in relation to sustainable consumption. The purchase of inexpensive products from low-wage countries, while profitable to consumers from high-wage countries, leads to their de-industrialization, with alarming socioeconomic consequences. If the objective of sustainable consumption is understood to include some social criteria we are left with difficulties in setting out priorities that should, more-over, be understood and supported by the population. Based on a survey carried out in Great Britain, Hobson (2002: 95) suggests an idea that can be connected to this challenge: 'This analysis argues that social justice, not sustainable lifestyles, has the most resonance with interviewees. As a result, not only do calls for rationalisation carry little cultural meaning, they also actively alienate individuals from the project of sustainable consumption.'

Discriminating between products, to identify those that meet sustainable consumption requirements, demands means of analysis lacking in consumers

The observations we made at the end of the previous point prompt us also, if we want to promote sustainable consumption beyond restricted circles where it may be rationalized, to study the way in which these subjects are repre-sented in the population as a whole. In this respect the discrimination between sustainable and not sustainable products is crucial. For social impacts of products difficulties will rapidly appear, outside of well-identified but limited initiatives, among which Fair Trade products (see several contributions in this volume), or in a more diluted way, products made in companies declaring meeting a variety of codes of conduct. But in ecological matters as well, field studies (Bartiaux, Chapter 7, and Rousseau and Bontinckx, Chapter 6, this volume) demonstrate the low level of identification in the population as a whole of more ecological products and behaviours, including those duly labelled. A collection of studies conducted in different countries show that attitudes to ecological problems vary in terms of problems and products, social categories, but also concrete local situations (Diamantopoulos et al., 2003). Aside from these technical obstacles which should be considered in the design of efficient policies and actions, there exists a fundamental criticism of the idea of placing the burden of responsibility for these choices on individuals (Dolan, 2002), introducing at the very least the importance of consumer group dynamics (Georg, 1999; Burgess, 2003), or other collective social drivers.

Attitudes in support of sustainable consumption are translated into only limited reductions of the negative impacts of consumption

This observation is commonplace in social science research on consumers in relation to the environment. 'The influence of environmental knowledge on environmental consciousness is small, effects of environmental knowledge and consciousness on behaviour are insignificant' (Kuckartz, 1995 quoted in Brand, 1997). In this volume, Bartiaux provides concrete examples showing that the higher social categories have both an 'environmentally friendly' attitude and greater impacts than social categories where the situation is the reverse, a result corroborated by different studies (Brand, 1997; Dobré, 2002). This is not surprising, since the most discriminatory factor, on average, of the increase in impacts is income level.

This does not mean that strengthening these attitudes, and providing information on possible actions (EC, 2004), is pointless. It is a pipe-dream, however, to count on this aspect in the hope of obtaining significantly different results within the pursuit of economic growth.

It seems more promising to gain better understanding of how consumers are partially 'locked in' to certain consumer behaviours owing to infrastructures (Green and Vergragt, 2002; Sanne, 2002; Southerton *et al.*, 2004) and the organizational networks in which their actions take place, in routine and ordinary forms of consumption (Shove, 2005). Consumers' concrete practices in managing their decisions and their time (Spaargaren, 1997; Cogoy, 1999) should also be studied.

From the standpoint of enterprises, at an ecological level the green profile of products is not a first order positioning factor, while at a social level voluntary commitments reflect limited changes

This observation should be related to two lines of reasoning that have been argued extensively in recent years. First, in a perspective of governance, it was considered that more changes in production patterns and products were to be expected from voluntary corporate initiatives. At the Johannesburg Summit in 2002, for example, this position was expressed both in the text of the Programme of Implementation and by the presentation of numerous partnerships with private enterprises. Second, from the earliest 'citizen-consumers' initiatives, it has been repeatedly claimed that consumers would be capable of influencing the market by their preferences for certain products that meet more ecological or social criteria.

This presumed model of influence nevertheless shows its limits in reality. Only a minority of products occupy 'ecological niches'. The 'green' argument is secondary in product profiling (Kong *et al.*, 2002). A survey made in Europe has demonstrated that consumer organizations and consumers are not among the top sources of influence in decisions for corporate environmental managers (Kestemont, 1999). Of course some campaigns of boycott have attracted attention and stimulated comments because of the new configurations and opportunities they could offer (e.g. Brent Spar, Clean Clothes Campaign). All in all, their impacts do not appear to deviate significantly from the general curse of world business.

Nevertheless, while there may be only a weak direct influence of 'market demands' within the meaning that concerns us, it is none the less true that codes of conduct and ratings of different sorts related to various aspects of sustainability are multiplying in the drive for corporate social responsibility (Livesey and Kearins, 2002). The influences favouring these changes bring consumers into the picture indirectly, though in a less important way than the influence of shareholders, company staff and certain stakeholders, acting in a perspective of limiting risk-taking in case of failure – risks whose level depends partly on regulatory decisions taken by the public authorities (e.g. legal liability in pollution cases).

These developments are leading to changes in certain corporate impacts and practices, whose effects seem limited on average. This does not prevent initiatives from being interesting and dynamic, however (see Fraselle and Scherer-Haynes, Chapter 13, this volume).

Possible changes in favour of sustainable consumption will have to involve the development of interactions with enterprises, a major challenge requiring that we move beyond superficial discourse on 'win–win' relations to shrewder observations. The Fair Trade initiatives already underway can also be studied in this perspective of interaction between producers and consumers (Le Velly, Chapter 14, Scherer-Haynes, Chapter 15, this volume), in particular in their possible instructive role, with a view to possible extensions in other forms in the market.

Information instruments are preferred tools for change even though their impact is weak

As we have seen, the pseudo-rational model – inspired by simple economic models – of the informed consumer changing his or her habits (and influencing the market) because he or she is informed is not very effective. Yet both a significant part of public policies and of the work of associations continue to count on it. This type of action is easier to introduce than structural changes through regulatory instruments or taxation, and is less likely to raise some of the contradictions that have been identified. What is also striking in the implementation of these instruments is the relatively low degree of professionalism in the profiling of such campaigns. They none the less have to face up to far more powerful and professional competitors, namely advertising campaigns for the sale of 'normal' products based on abundant and costly market studies,[5] and aimed at well-defined targets. In this new version of David versus Goliath, the ethical potential of calls for sustainability does not protect them from a certain inefficiency. In addition, even advertisers with heightened awareness prove incapable of coming up with messages to encourage limits on consumption, which would contradict the broad standards of their profession.

The fact remains, however, that progress is possible in terms of making the promotion of certain products, such as Fair Trade goods, more professional (see De Pelsmacker *et al.*, Chapter 8, this volume). Moreover, the substance of messages in support of sustainability plays an indirect role in supporting various initiatives.

The objectives of a sustainable consumption policy are relatively vague

The offspring of sustainable development, sustainable consumption, has the same ambiguous relations in the combination of ecological and social questions. Environmental criteria are dominant in relevant programmes, while social criteria appear erratically, and certain subjects like Fair Trade or corporate social responsibility are recurrent. The economic criterion is generally implicitly integrated.

In the last EEA report on households (2005), for instance, sustainable consumption was studied solely from the angle of the reduction of ecological impacts, with some corrections to avoid burdens from this policy that would affect disadvantaged people. The Integrated Product Policy, the focus of European hopes in this field for the past few years (Oosterhuis *et al.*, 1996), is taking a long time to materialize (Rubik and Scholl, 2002) and bring significant results (Pallemaerts *et al.*, 2006). In international reports promoting sustainable consumption and production, like that of the OECD or the Johannesburg Plan of Implementation, underestimation of social aspects is seen in particular in the fact that the implications of planned programmes on employment are barely addressed. Another example: over and above limited initiatives to promote Fair Trade, what is the connection with structural changes to promote greater fairness in trade in general? While these questions are not absent from the preparation of a programme to change production and consumption patterns in the wake of Johannesburg (see above), the objectives and means are muddled, with few priorities. What we can say for now is that, in programmes meant to address this issue, sustainable consumption is not being analysed with coherent objectives from the social, environmental and economic perspectives. This situation results partly from the contradictions mentioned above, but these should not prevent us from seeking more coherent policies that are targeted, limited and differentiated.

Concerning environmental policies alone, not all objectives are defined either. According to a simple but useful distinction introduced by Sachs (1999a), environmental policies have traditionally dealt with questions of 'nanograms' (typically of different forms of chemical pollution) but there is a need to switch to objectives covering 'tonnes' (massive quantities of resources, typically of decoupling and dematerialization objectives). However, this second type of objective, which calls for more massive transformations, needs more solid foundations differentiated in terms of vectors, in addition to the reduction of greenhouse gas effects, the priority of priorities. This is not the case at present.

Thus, change is needed in the profiling of effective sustainable consumption policies and goals, taking account of experiences accumulated in these fields for over a decade (Zaccaï, 2006).

Although the goal of sustainable consumption raises many questions, the issue appears consensual rather than being the subject of policy debates

This characteristic also appears subsequent to what has taken place in recent years with regard to the more general objective of sustainable development. Sustainable development appears to be a broad principle which is interpreted and referred to, but its implementation as a general project is being pursued in different types of plans and programmes which only marginally steer the actions of governments (for the situation at European level see Niestroy, 2005). Working to change production and consumption patterns is seen as a main way of implementing these objectives, but this aim is apparently not being achieved.

Of course, there have been numerous policy debates on the different issues that emerge from this objective, such as the reduction of greenhouse gas emissions, taxation on several polluting products, or changes in international trade regulations to name a few. In these contexts, certain arguments or objectives related to sustainable consumption are brought up, but this concept lacks the capacity to serve as a unifying argument. The internal contradictions we have identified contribute in various ways to this state of affairs. Another explanation, which is related, is the low take-up of the concept of sustainable consumption by the public.

It is hard to guess how this concept will evolve in the future. One possibility is that it will be maintained as a pool of arguments and analyses, relatively dissociated from concrete political achievements. In contrast, if it is to inspire decisions more directly, it is obvious that today's limp consensus cannot be maintained and that choices will have to shed light on important internal contradictions. As Pignarre put it in another context (public health choices in the face of the pharmaceutical industry), 'bringing a problem into politics' implies tying it in to existing practices and knowledge, which will also change the way the problem is raised because 'such a framework cannot be found single-handedly. Once a political issue starts taking on importance and is championed by a group of concerned persons, it branches out, becomes richer and more complex – in a positive sense. Its supporters meet other persons or groups, mobilise them and transform them in turn' (Pignarre, 2004: 174).

In this perspective perhaps one contribution of research would be to offer clarification and analyses of different aspects of this goal of sustainable consumption, which today is formulated in policies in ways that are at the same time too rational, partial and abstract in certain regards. The ways of resolving these contradictions are not yet clear, but historically these challenges are only in one stage of several and will inevitably evolve. In this context, the role of consumers constitutes a particularly rich question, insofar as it links psychology and sociology, values and actions, individuals and political organization, and ordinary life and an ideal for bringing about change.

Introduction to the contributions

Generally, sustainable consumption, behind a thin veil of developmental considerations, means ecological sustainability. This book set out to explore a wider meaning of the theme, inspired by the full sense that can be given to sustainable development. Thus Part I 'Consumption: what kind of a problem for sustainable development?', with chapters interrogating the criteria and hence the approaches that may be used to assess sustainable consumption today.

Paul-Marie Boulanger opens this topic by introducing the concepts of *misconsumption* differing from *overconsumption*. Backing up his analysis with economics and moral philosophy, the author discusses some moral implications of consumption, a topic that we will find again in a number of subsequent chapters. He then presents aggregated quantitative results on overconsumption in different regions of the world, aiming to assess the sustainability of national levels of consumption.

Chapters 3 and 4 study first the ecological criteria and then the relations between consumption and well-being. Anton Schoot Uiterkamp, who co-edited a book which was a precursor on this subject in Europe (Noorman and Uiterkamp, 1998), offers an analysis based on European programmes which have studied actions that may be carried out in household consumption. He points out, in particular for energy matters, the extent of the challenge for which we have only limited solutions thus far. The subjective contribution of consumption to welfare is then examined by John Lintott via theoretical analyses and survey results. The pursuit of differential standards relative to the individual's group memberships appears to be a decisive factor, which brings us to the severe limits on change through individual actions and prompts a drive to seek collective action, a key theme running through a number of contributions.

Closing this first part, Grégoire Wallenborn strives to introduce different ways in which sustainable consumption will be studied as well as the choices that will be made in terms of models influencing the political approaches to it. His conclusion is that the more complex the model and the more it integrates the multiple characteristics of the act of consuming, the harder it is to identify the factors for changing consumption patterns.

Part II offers insights into a central dimension of this work: 'Who is sensitive to sustainable consumption, and why?' Five contributors examine how consumers relate to several issues such as the ecological components of impacts, Fair Trade, or the pursuit of various social ambitions and so on. Dealing sometimes with the population as a whole where responsible consumption is rare, and sometimes with limited groups where it may be analysed in greater depth, these chapters provide data and typologies on consumers' knowledge, values, motivations and changing behaviours.

In this and Part III, analysis relating to Fair Trade will be very evident. Aside from the ecological criteria, Fair Trade appears to be increasingly associated with sustainable development by the actors. It is a rich field in which to collect in-depth analysis about one aspect of social criteria asociated with consumption: 'fairness' in its trading, and it can document comparisons between some social and some ecological motivations within the wide approach taken towards sustainable consumption.

Catherine Rousseau and Christian Bontinckx, basing their findings on field research using an original psychosocial model, will highlight the low level of ecological motivation per se in taking action, and provide tools and recommendations for differentiated and collective actions with a view to bringing about effective changes. Françoise Bartiaux paints a picture that confirms the weak links between ecological knowledge and action in a population, a result which remains underestimated in traditional campaigns based on the model of informing the consumer about certain impacts in the hope of making him or her more selective in his or her 'rational' choices. Coline Ruwet elaborates original typologies of attitudes and patterns of action on the basis of a field study, this time focused on consumers asserting their commitment. This research enters into the justifications claimed by the subjects, while placing them in sociological analyses that go beyond responsible consumption to examine how individuals relate to social production systems. The contribution by Gautier Pirotte offers analogies with Ruwet's study, concentrating on consumers who support Fair Trade. Here too, social representations, in this case on questions of equity and North–South solidarity, are decisive. For their part, Patrick De Pelsmacker, Wim Janssens, Catherine Mielants and Ellen Sterckx present original results from market analysis methods applied to Fair Trade. This chapter provides an interesting counterpoint to the sociological analysis of Pirotte, sharing similar ground. One of the measures to promote Fair Trade recommended by De Pelsmacker *et al.* is the wider availability of Fair Trade products in mass distribution. It is worth noting that in matters of ecological products, Rousseau and Bontinckx made a similar recommendation for the multiplication and availability of supply.

The main field of research where these five contributions originate is Belgium, but in several cases the results are compared to other findings in the literature, and are always put into perspective to draw conclusions connected to the key questions of the book. All in all, we see the portrait of various motivations towards sustainable consumption progressively acquiring more colour and detail, raising the question of the impacts of such (limited) actions of consumers on the general curse of society.

The final part of the book, 'Dynamics of sustainable consumption', contains contributions studying various dimensions of the interactions between consumers' actions and their impacts on several social processes. Some of them advocate paths for changes.

First, Michelle Dobré elaborates an original figure of 'resistance' through some practices of ordinary consumption. Her analysis is backed by sociological history, results from focus groups and statistical elements (with the wise question: How to observe what is *not* consumed?). The analyses of Dobré make the link between the previous part (what are the motivations of consumers?) and the place of these dynamics in society a subject which is thoroughly investigated by Serge Latouche. This author is a fervent advocate for a 'de-growth' economy in France, after having elaborated radical critiques of development in several books (2003, 2004). His chapter makes the case of a militant position in favour of alternative actions of citizen-consumers that will lead them, in many aspects, outside of the market as it stands today. In contrast to this radical position, the analysis of Nadine Fraselle and Isabelle Scherer-Haynes outlines the profile of a reformist transformation, where the role of enterprises has to be very important, if not central. The authors are reflecting on the characteristics of a more 'social economy' and on the organized drivers for change among stakeholders in society. This final part closes with two case studies allowing the reader to discover some of these dynamics of change in the field of Fair Trade in greater detail. Ronan Le Velly exposes the tensions between distinct objectives included in the Fair Trade sector, such as on one side working in favour of excluded populations, and on the other meeting the commercial standards asked by demanding consumers. This analysis sheds light on the conditions by which an extension of Fair Trade can be expected, one of the key questions in this domain (Goodman, 2005). Another key question is the extent to which Fair Trade effectively brings improvements to the lives of Southern producers. Thanks to recent field research among Indian cotton workers, Isabelle Scherer-Haynes documents the reality of the transactions of norms along the whole chain from consumers back to producers.

In the end, one of the teachings we may get from all these results and analyses is that sustainable consumption is not only about sustainability, it is perhaps even more about *consumption* (not forgetting production and trade). We learn elements about sustainable consumption motivations and, crucially, compare them to the extent of changes needed for a more ecological and fair world. But what we learn is raising many fundamental questions about consumption in general, what it means, and what are its powers for the individual and for society. In a brief conclusion, we return, together with Paul-Marie Boulanger, to a profile of sustainable consumption today and to its moral and political conditions.

Before letting the reader elaborate his or her own idea of what will remain of the contradictions underlined in the opening of this introduction, on behalf of all the contributors, I would like to express our gratitude to the Belgian Science Policy for its generous financial support of this project. The preparatory seminars that could be organized within this framework were interesting for all the research fellows who participated, enabling them collectively to sort out the intricacies of the subject. We feel sorry that we could not publish all the interventions exposed on these occasions, and especially the works of Eric De Keuleneer, Charles Maier, Michael Goodman, Tim Jackson, Pierre Stassart and

Melanie Louviaux. Within the BSP, Marie-Carmen Bex has always shown her active interest in this project and facilitated, with a smile, its administrative dimensions; many thanks for that. Thanks also to Tom Bauler and Coline Ruwet who helped me in some key editing tasks. And to Celine Curvers who, in the meantime gave birth to her Anton, and mostly to the high-level secretarial help of Vanessa Demeuldre for formatting the manuscript, also with a smile. Thank you also to Jonathan Burnham across the Atlantic, and to Gabrielle Leyden in Brussels, for translating and correcting some of the contributions. To all my partners in this project, it has been a chance for me to discuss and work with you, and this will stay in my mind as a cheerful experience in my professional life.

Notes

1   *10-Year Framework on Sustainable and Consumption (The Marrakech Process)*, Division for Sustainable Development (DESA), United Nations, and United Nations Environment Program (UNEP). See also the Oslo Declaration, signed by more than 250 scientists in 2005, and pointing at the severe implementation gap relating to this framework, three years after its launch (http://www.oslodeclaration.org).
2   The Belgian Science Policy set up a programme of support for sustainable development policies which funded the project from which this book originated, through a series of seminars held in Belgium in 2004 to 2005.
3   At the 'Rio+5' UN Conference in 1997, based on European proposals, references to eco-efficiency ('Factor 4 and 10') were deleted.
4   According to China's environment administration, 79 per cent of total air pollution in 2005 will come from vehicle gas emissions. In the country's fourteen biggest cities, air pollution is thought to kill 50,000 newborns yearly and to be the cause of 400,000 cases of respiratory illness. In this context, by adopting the Euro III European standards (which will apply from July 2007) and by programming Euro IV for 2010, China plans to reduce polluting vehicle emissions on its territory by 30 per cent and subsequently by 60 per cent. The 'rebound effect' is nevertheless still a topical question since the country could move from 30 million passenger cars today to 131 million in 2020, with the corresponding consumption of petrol, to say nothing of the fact that the trend today is to disparage bicycles and show a preference for cars. Based on Alter Business News (94) and SEPA, Chinese Environmental Protection Agency.
5   A striking observation in this connection is provided by UNDP: annual global advertising expenses represent about ten times the amount needed to eradicate hunger, offer access to drinking water for all human beings, provide them with decent housing and combat major epidemics (quoted in Viveret, 2002: 39).

Part I

Consumption

What kind of problem for sustainable development?

2 What's wrong with consumption for sustainable development

Overconsumption, underconsumption, misconsumption?

Paul-Marie Boulanger

Introduction: consumption assessment and sustainable development

Even though the Rio Conference and Agenda 21 have given a new impetus to the critical assessment of industrial societies' consumption patterns (Zaccaï, 2003), it is by no means something new in the intellectual heritage of modernity. Indeed, critiques of consumption, consumerism and the consumer society have been pervasive and widespread from the very beginning of the modern era. Nearly all of the most lucid observers of the nascent consumer society were at least doubtful about the real benefits, both for society and the individual, of the emerging cultural models and behaviours regarding consumption. Rousseau,[1] Marx, Tocqueville, Durkheim, for instance, have all – even if with varying eagerness – manifested some apprehension concerning the growing importance of consumption in modern society. All of these classical objections are based on a particular conception of the 'good life' (see Dobré, Chapter 11, this volume). It is in the name of a supposed 'better', more authentic and valuable conception of life that they despised the evolution of civilization and society. Basically, they argue that consuming has become deceiving for the consumer. Hence they vindicate the consumer society from a consumer point of view, not from the point of view of those who are *not invited* to the banquet.

With sustainable development, critiques of consumerism have found a new, extra rationale: its environmental requirements and impacts are such that it cannot be generalized through space and time. In short, it is unsustainable. However, for many champions of sustainable development, unsustainability does not seem to make the more traditional critiques of consumption obsolete or superfluous. There is thus an apparent contradiction[2] here: if consumerism is so flawed, why care about its sustainability? And if it is unsustainable, what is the point of objecting to it?

But sustainability is not just a matter of fact; it is fundamentally an ethical problem. What is wrong with unsustainability is not so much that my way of life cannot be generalized to others (after all they are not compelled to want it), but that it makes it probably impossible for others, now and later, simply to live a decent human life. Therefore, in what follows, I make a distinction between critiques of consumption based on different conceptions of the good life – that is, from a consumer point of view – and critiques based on some conception of justice. The distinction is necessary, I think, because if objections to consumption, from a consumer point-of-view, are certainly relevant and have to be taken into consideration in an overall assessment of our way of life, they are insufficient to legitimate pressing demands for more responsible behaviour from the consumer ('Live simply so others may simply live' (M. Ghandi)) or public interventions for fostering such behaviour. In my opinion, these must be based not on – necessarily relative – conceptions of the good life but on ethical reasons, i.e. on arguments about justice and fairness.

Critiques of consumption

The misconsumption stance

Many of the objections to consumption may be referred to one or another of the concepts and relations depicted in Figure 2.1, which I borrow from Jackson and Michaelis (2003: 17),[3] with just a few additions. It synthesizes the most widespread conception of the relationships between production, consumption, needs, well-being, society and the environment.

As is apparent in the figure, production is supposed to be a response to a demand originating in consumers who seek to maximize their welfare through consumption of goods and services necessary to the satisfaction of their needs or wants. However, wants (or needs) being intrinsically insatiable,[4] the quest for satisfaction is unending, leading to demand of new products and services and therefore to more and more production. The arrow between demand and production in Figure 2.1, and the lack of feedback between the two, indicates that the consumer is perceived as the deus ex machina of the overall economic process. Of course, many critiques of capitalism have denied this intensely and argued that, far from being the master of the play, the consumer was more like a puppet on a string or a pawn on the chessboard, manipulated by the capitalist class or by the 'system'.

Marx is probably the first to argue that in capitalism the consumer, far from being free, let alone sovereign, is deeply alienated because:

> Every person speculates on creating a new need in another, so as to drive him to fresh sacrifice, to place him in a new dependence and to seduce him into a new mode of enjoyment and therefore economic ruin. Each tries to establish over the other an alien power, so as thereby to find satisfaction of his own selfish need. The increase in the quantity of objects is therefore accompanied by an extension of the realm of the alien powers to which man is subjected, and every new product represents a new potentiality of mutual swindling and mutual plundering. Man becomes poorer as man, his need for money becomes ever greater if he wants to master the hostile power.
> (Marx, 1844: 306)

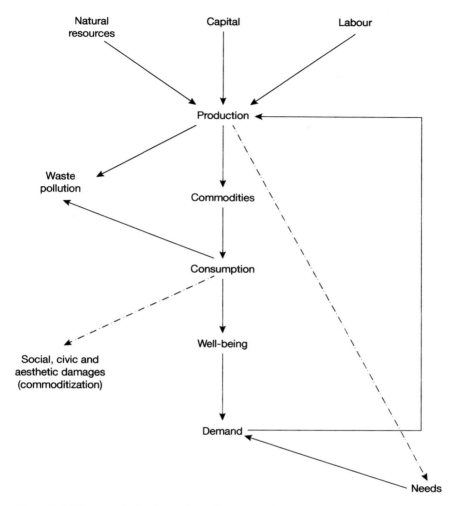

Figure 2.1 The standard view of production and consumption (with some non-standard additions)

Source: Jackson and Michaelis (2003: 17).

Becoming 'poorer as man' means becoming the slave of inauthentic needs imposed on him by others and being more and more passive; that is, as Elster (1985: 80) put it: 'directed towards passive consumption rather than active creation'. Of course, man enjoys consuming and may have the illusion of being free in his consuming behaviour, but:

> The fact that the labourer consumes his means of subsistence for his own purposes, and not to please the capitalist, has no bearing on the matter. The

consumption of food by a beast of burden is none the less a necessary factor in the process of production, because the beast enjoys what it eats.[5]

(Marx, 1867: 627)

The evolution of capitalism during the twentieth century, with its extraordinary expansion of marketing and advertising,[6] was seen by many as a striking confirmation of Marx's early diagnostic. It is especially the case for Marcuse, who elaborated further on the distinction between real and false needs.[7]

We may distinguish both true and false needs. 'False' are those which are superimposed upon the individual by particular social interests in his repression: the needs which perpetuate toil, aggressiveness, misery, and injustice. Their satisfaction might be most gratifying to the individual, but this happiness is not a condition, which has to be maintained and protected if it serves to arrest the development of the ability (his own and others) to recognize the disease of the whole and grasp the chances of curing the disease. The result then is euphoria in unhappiness. Most of the prevailing needs to relax, to have fun, to behave and consume in accordance with the advertisements, to love and hate what others love and hate, belong to this category of false need.

(Marcuse, 1964: 2)

Of course, as Marcuse himself admits, it is up to the individuals themselves to discriminate between their real needs and the ones imposed on them by advanced capitalism. Yet they are unable to do so because of the ideological climate into which they are plunged (Marcuse, 1964: 2).

In the final analysis, the question of what are true and false needs must be answered by the individuals themselves, but only in the final analysis; that is, if and when they are free to give their own answer. So long as they are kept incapable of being autonomous, as long as they are indoctrinated and manipulated (down to their very instincts), their answer to this question cannot be taken as their own. By the same token, however, no tribunal can justly arrogate to itself the right to decide which needs should be developed and satisfied. Any such tribunal is reprehensible, although our revulsion does not do away with the question: How can the people who have been the object of effective and productive domination by themselves create the conditions of freedom?

Since then, the argument of the manipulated consumer has been restated by writers like Galbraith or, still more recently, by Miles:

Not only does consumerism structure our everyday lives, but it does so by offering us the illusion of consumer freedom when, at least, to a certain extent, such freedoms are inevitably constructed and constrained.

(Miles, 1998: 5)

One may add that the consumer is generally also a worker or a shareholder and has, as such, a vested interest in the development of production, his or her income (profits and/or wages) depending on the expansion of the economy. This may be considered as another form of alienation, understood in its clinical signification of mental illness, as a kind of schizophrenia: Dr Jekyll the consumer and Mr Hyde the producer (or the other way around?).

There is another version of the objection to the myth of the sovereign consumer: the thesis of the consumer *lock-in*. The problem, it is argued, is not so much marketing or advertising but the socioeconomical functioning of society, which requires high levels of income and consumption just to meet fundamental economic needs (Sanne, 2002).

Segal, for instance, denies that the high level of consumption in the First World means that essential needs of the population are met:

> Rather, I will argue, most Americans legitimately feel hard-pressed economically. Fundamental economic needs are either unmet or can be met only at high levels of income and consumption. This, however, is not the result of some fixed universal feature of the human condition but, rather, emerges from the socio-economic conditions of a particular society at a particular point in time, and, most importantly, from how that society decides to respond to economic needs.
>
> (Segal, 1998: 176)

He shows that the real cost of satisfying fixed needs has grown substantially over past decades. Actually, this is something already observed in 1973 by Richard Wilkinson for which it has been true since the very beginning of the Industrial Revolution:

> Society exchanged poverty in one sphere for poverty in others which seemed less vital. Problems of transport, urban sanitation, entertainment, education and social activity became increasingly pressing. It is to the development of problems such as those that we owe much of the modern urgency for consumption.
>
> (Wilkinson, 1973: 175)

These problems, in turn, are created by new working conditions totally different from those of pre-industrial societies:

> The type of work which the population of an industrial society must perform has several fairly obvious effects on their needs as consumers. People have to adapt themselves physically, mentally and environmentally to meet the demands which their work imposes on them.
>
> (Wilkinson, 1973: 177–178)

All in all, to go back to Figure 2.1, the critique against the sovereign consumer may be synthesized by drawing an arrow between 'production' (or, perhaps, capital?) and 'needs', showing that the consumers' wants, far from being the manifestation of their free will, are shaped by the needs of production, i.e. of the 'system'.

Consumption and well-being

Although in the standard view more consumption necessarily leads to more well-being, this is highly controversial. Of course, if there is no more in well-being than utility, as in utilitarianism, the proposition is plain tautology: more utility brings more utility. Moreover, there is no point in disputing the fact that if consumers do consume, it must be because it contributes to their well-being: who knows better than the consumers themselves what is good for them? Where their well-being is and how to achieve it?

This reasoning has been challenged from three different perspectives:

- The first perspective is somewhat philosophical and may be associated with the names of Sen, Nussbaum, Crocker and so on. They oppose utilitarianism with an Aristotelian conception of well-being based on the couplet 'capabilities–functioning'; that is, on the real freedom to be and do what one has good (informed) reasons to value. Although this may appear rather scholastic, it has very concrete implications in terms of assessment, well-being indicators and development policies. For instance, while utilitarianism legitimates gross domestic product (GDP) as the best indicator of development, a 'capabilities–functioning' point of view calls for something resembling the Human Development Index.
- The second perspective is empirical and subjective: it is based on the observation that past a given level of affluence, more consumption does not bring more happiness or well-being, as was demonstrated by Easterlin (1995). Thus there would not only be external, environmental limits to consumption but internal ones as well. I will not dwell further on this topic, as John Lintott handles it thoroughly in a section of his chapter.[8]
- The third perspective has to do with 'consumption efficiency' and what Segal calls the need-required income (NRI), which is the level of income necessary to meet the needs of the population. The NRI, according to Segal, is constantly growing, but this does not mean that there is any betterment in the well-being of the population. For instance, the per capita consumption level of America in 1994 is twice that of 1960 and three times that of 1929. It is doubtful that well-being has grown in the same proportion. Actually, it hasn't, as Lane (2002) and Layard (2005), among many others, demonstrate. This is not without consequences in terms of prospects of voluntary consumption reduction. In fact, it means that there is not much room for reduction without real welfare losses. Therefore, it should be a 'central political objective' to prevent the NRI from rising.

Consumption and the community

A relationship between consumption and community, lacking in the original version of Figure 2.1, has been added here because it can be problematic from a sustainable developmental point of view.

For instance, consumerism correlates with the replacement of the small, local independent retailers by large-scale (national or even multinational) corporate chains. The consequences of these changes on the vitality and diversity of local and community life are pervasive. The small retail business was generally localized in the centre of the city where it contributed to the animation and security of the street (Jacobs, 1961), all kinds of public goods that were given for free and for which it is now necessary to pay. On the contrary, the megastores are usually located on the periphery of towns in unpopulated areas, which become insecure after the shops' closure at night. In addition, this strengthens car dependency and traffic problems.

Moreover, on aesthetic and sociological grounds, some deplore the '*McDonald-ization*' (Ritzer, 1998) of society as homogenization of neighbourhoods, landscapes and lifestyles.

Finally, consumerism is also sometimes blamed for its supposed harmful impact on social capital, which is increasingly acknowledged as an indispensable productive asset. This has something to do with *commoditization*, which Manno defines as:

> The tendency to preferentially develop things most suited to function as commodities – things with qualities that facilitate buying and selling – as the answer to each and every type of human want and need.
>
> (Manno, 2002: 70)

Not only does this process of commoditization undermine the social fabric by making marketable satisfiers that were beforehand provided only by non-economical institutions such as family and kinship, neighbourhood, communities, and so on, it also has negative impacts on the material and energy requirements of consumption insofar as it tends to substitute more resources demanding satisfiers to less demanding ones.

Summary

All the above-mentioned objections, even those related to communities, neighbourhoods and social capital, may be characterized as consumer-oriented (Everett, 2001). Here, consumption is considered perverse insofar as, while it is supposed to be pursued for the consumer's sake, it results in deception and welfare losses. This is why I suggest calling it '*misconsumption*'. It is consumption assessed from the consumer point of view against values and norms such as authenticity, health, freedom, welfare, happiness and other ingredients of the 'good life'. It conveys the idea that one's consumption does not match one's authentic needs, wants, intentions, interests, well-being and so on.

Misconsumption critiques are generally from the rich, First World's consumer point of view. Admittedly, one could find it indecent to complain about consumption, consumerism, the consumer society and the like when none of us would agree to exchange his or her life of harassed consumer with that of the average Third World inhabitant. However, insofar as – willing it or not – our way of life is given – or perceived – as the ultimate goal of 'development', if not as the end of history (Fukuyama, 1992), it is of the utmost importance to assess it objectively, including with respect to consumption. Indeed, if sustainable development is a development that can be generalized in time and across space, the question we ought to ask ourselves is not only if our development can be generalized but also, and first, *if it should be*. This is why critiques of misconsumption are relevant for sustainable development.

Thus far, nothing has been said about the relationship introduced earlier in Figure 2.1 between production, consumption and the environment, such as waste disposal, pollution and material and energy requirements of consumption. This does not mean that it cannot also be assessed from the consumer point of view and categorized as misconsumption. I think this is indeed the case if the environment is considered only for the sake of consumers' health, quality of life, and so on. When it is considered from the others', the non-consumers' point of view, it is more accurate to classify it in the overconsumption category.

The overconsumption stance

Overconsumption is another oriented objection; it is consumption assessed from the non-consumer's point of view, for instance, the current poor or future generations.[9]

When and why can one speak of overconsumption? Certainly it is consumption in excess but in excess of what and why in excess? There are two issues here; the first is 'how much is too much?';[10] the second, 'what's the problem with it?' I think it is impossible to tackle the first without reference to something like 'consumption norms', hence some definition of needs. Concerning the second issue, and putting aside considerations stemming from the (over)-consumer point of view, I propose to qualify as overconsumption (of something by someone) only that part of consumption which is above a given consumption 'norm', such that it makes it unlikely for others to achieve a 'normal' consumption.

In other words, there is overconsumption if and only if it results in underconsumption elsewhere (contemporaries) or later (future generations).

Therefore, from a sustainability perspective, there is overconsumption when:

- Some people do not have access to sufficient amounts (i.e. above a specified threshold or norm) of a given resource or of resources in general (underconsumption);
- Others enjoy levels of consumption of these resources that lie above the aforementioned threshold (overconsumption as such);

- There is a causal relation between the deprivation of the former and the (over)consumption of the latter.

The third condition is indispensable from a sustainability point of view: overconsumption is a concern if and only if it is harmful to others.

Overconsumption as excessive appropriation of common resources

There can be no overconsumption of non-rivalrous non-excludable goods such as security, air, love, information and so on. Someone surfing on the web night and day may very well be misconsuming but certainly not overconsuming (except perhaps for some overload effect). Only resources existing in limited supply can be overconsumed. Nor is there overconsumption if my consumption behaviour has no effect on someone else, as was stated earlier. In other words, I may be misconsuming my own resources, but not overconsuming them. I can only overconsume resources over which I do not have absolute property rights.

This is clearly the case with the Earth's biophysical resources that human-kind[11] possesses in common. If non-renewable they are limited in an absolute sense; if renewable they may be used up to their renewal rate. At what conditions can these resources be appropriated?

For Locke (1690), everyone has a natural right to appropriate[12] them by

> '*mixing* one's labor' with natural common resources, provided 'there is enough and as good left in common for others'.
>
> (Locke, 1966, ch. V: 15)

This is known as the Lockean proviso. Three qualifiers are important: '*enough*', '*as good*' and '*in common*'. The first insists on quantity, the second on quality (and substitutability) and the third on accessibility. Thus, from the Lockean perspective, there is overconsumption if the consumption of common resources by some is such that not enough of them or of others as good are left in common for others.

Admittedly, this proviso raises some difficulties, both empirical and philosophical, some of them already identified by Spencer:

> Further difficulties are suggested by the qualification, that the claim to any article of property thus obtained is valid only 'when there is enough and as good left in common for others'. A condition like this gives birth to such a host of queries, doubts, and limitations, as practically to neutralize the general proposition entirely. It may be asked, for example – How is it to be known that enough is 'left in common for others'? Who can determine whether what remains is 'as good' as what is taken? How if the remnant is less accessible? If there is not enough 'left in common for others', how must the right of appropriation be exercised? Why, in such case, does the mixing of labour with the acquired object cease to 'exclude the common right of

other men'? Supposing enough to be attainable, but not all equally good, by what rule must each man choose?

(Spencer, 1851: 127–128)

These difficulties are the very subject matter of a science of sustainable development. Other difficulties – notably the risk of *reductio ad absurdum* – have been identified and discussed very thoroughly by R. Nozick (1974). He suggests interpreting the proviso in a wider sense, as the obligation not to make others (the non-appropriators) worse off from my appropriation, and therefore to compensate them for the lost resources or opportunities. This interpretation is based on what is known as the second Lockean proviso, correcting or qualifying the first, which states that:

> The common ground rules of human coexistence may be changed only if everyone can rationally consent to the alteration. . . . He [Locke] claims that the lifting of the enough-as-good constraint through the general acceptance of money . . . satisfies a second-order proviso.
>
> (Pogge, 1998: 508)

In other words, as soon as one 'consents to the use of money' there could be nothing unjust in the unilateral and discretionary appropriation of a disproportionate share of common resources.

One may wonder if it is really the case today that the rich cause harm to the poor by appropriating, without sufficient compensation, an excessive share of the Earth's resources. After all, environmental goods are sold and purchased on international markets, not extorted or plundered. However, even if from a macroeconomic point of view the Southern States were rightly compensated (by the North's consumers), it would not necessarily follow that the South's poor benefit from the compensation. As Pogge put it:

> Many of the global poor are today just about as badly off, economically, as human beings could be while still alive. It is not true, then, that all strata of humankind, and the poorest, in particular, are better off with universal rights to unilateral appropriation and pollution . . . our world, therefore, does not meet the requirements for lifting of the Lockean proviso. The exclusion of the poor from a proportionate share of resources therefore manifests an injustice that citizens of the affluent states and the 'elites' of the poor impose by force. Accepting this exclusion voluntarily would be rational for the global poor only if they were compensated for it by being effectively guaranteed an adequate share of the benefits that otherwise derive from their unilateral appropriations.
>
> (Pogge, 1998: 508–509)

Those who appropriate an excessive share of natural resources are not only First World consumers but also the Third World's ruling classes such as the Saudi royal family or, for example, what Pogge calls the 'Nigerian kleptocraty'.

However, it is not the case that environmental goods are adequately priced on markets. Chichilnisky (1994) showed that inadequate definition and enforcement of property rights are responsible for a general underpricing of environmental goods leading to a form of subsidization of the North's consumers by the South. This point has recently been re-established by leading economists such as Arrow, Dasgupta and Mäler, among others:

> owing to ineffectual systems of property rights, natural capital is frequently underpriced in the market: in extreme cases such capital is free. In calling a system of property rights ineffectual, we mean that those inflicting damage (e.g. destroying mangroves in order to create shrimp farms; logging in the uplands of watersheds) are not required to compensate those who suffer the damage (e.g. local fishermen dependent on the mangroves; farmers and fishermen in the lowlands of the watersheds). Even where property rights are reasonably well-defined, they are often unenforced. This suggests the possibility that countries that are exporting resource-based products (they are often among the poorest) may be subsidizing the consumption of those countries that are importing these products (they are often among the richest). . . . High levels of consumption in rich countries may promote excessive resource degradation in poor countries, which imperils well-being in the poor countries.
>
> (Arrow *et al.*, 2002: 18)

Externalities, market imperfections or unenforced property rights are ostensibly neutral and objective ways to call what others, without their tongues in check, call 'ecologically unequal exchange' (Martinez-Alier, 2002): asymmetric power relations and plain exploitation.

To summarize, from an appropriation standpoint, there definitely is overconsumption of natural resources by some (the rich, mainly in the North but elsewhere also) leading to underconsumption for the many who do not get their fair share of these resources and/or are not adequately compensated for their loss. But the problem is not overconsumption as such. The problem is that there is a causal link between overconsumption and poverty. North's (and South's riches) overconsumption is made possible notably because of the underpricing of the South's environmental resources leading to their overexploitation without really contributing to help South's poor escaping poverty.[13]

Overconsumption as insufficient savings

In the preceding subsection, consumption was understood in its ecological meaning, as transformation by humans of low entropy resources into high entropy ones, or as any human activities that interact with and alter the biophysical world.[14]

For an economist, such a definition hides the very important distinction between consumption and investment. Indeed, from an economic point of view, not all of what is given by nature or built by man is necessarily consumed. A part of it might (and ought to) be saved and invested in order to guarantee future consumption.

Therefore, when an economist asks the question: 'Are we consuming too much?', what they have in mind is: 'Are we saving enough to maintain our productive capacity and guarantee our own future welfare and that of our successors?'[15] In other words, does consumption today jeopardize consumption tomorrow?

Saving and investing consists in maintaining or improving the productive assets indispensable for generating future income and consumption. These assets may be classified as natural (the Earth's biophysical resources), produced (techno-economic capital), human (workforce skills, talents, know-how) and social (social norms and values that foster cooperation, trust, and so on).[16] The World Bank (Hamilton and Clemens, 2000) has worked out a measure of savings that more or less takes into account all these kinds of assets;[17] that is, not only man-made (economic) capital as usual but also human and natural capital.[18] It is called 'genuine saving' and consists in adding to and subtracting from the gross domestic saving as computed in the usual national accounts some investments or depreciations relative to human capital and the environment. More precisely it is computed as follows:

Genuine savings:

- = Gross domestic savings
- − Consumption of fixed capital
- + Education expenditure
- − Energy depletion
- − Mineral depletion
- − Net forest depletion
- − Carbon dioxide damage.

Table 2.1 shows genuine savings for some OECD and non-OECD countries. The genuine saving is supposed to be a sustainable development indicator. A positive value should be interpreted as signifying that the country is on a sustainable development path; a negative value points to a path that is not so.

Our purpose here is not to discuss the methodology and foundations of such an indicator, but to pinpoint some of its very counter-intuitive, if not repugnant, conclusions. First, as Table 2.1 shows, countries such as Belgium or the United Kingdom have a 0.0 value for mineral depletion. This is justified by the fact that there is no significant mining activity in these countries. However, it does not mean that the economic activities of Belgium or the United Kingdom are so 'dematerialized' as to incorporate no minerals at all. In fact, both countries probably consume more minerals and energy per capita than do Brazil or India. But, insofar as they do not come from domestic mines or energy

Table 2.1 Genuine savings for selected countries (1997)

Country	Gross domestic savings	Consumption of fixed capital	Net domestic savings	Education expenditures	Energy depletion	Mineral depletion	Net forest depletion	Carbon dioxide damage	Genuine domestic savings per capita ($)
	% GDP 1997	% GDP 1997	% GDP 1997	% GDP 1997	% GDP 1997	% GDP 1997	% GDP 1997	% GDP 1997	
Australia	22.9	13.7	9.2	5.3	0.6	1.3	0.0	0.4	-37
Belgium	25.1	9.9	15.2	3.1	0.0	0.0	0.0	0.2	4708
Brazil	18.6	10.4	8.2	4.7	0.7	0.6	0.0	0.2	-157
Canada	22.3	12.3	10.1	6.3	2.9	0.2	0.0	0.4	623
India	19.9	9.4	10.4	3.3	2.3	0.4	1.6	1.4	-24
Japan	30.3	16.0	14.3	4.7	0.0	0.0	0.0	0.2	6587
Mexico	25.8	10.0	15.8	4.4	5.4	0.2	0.0	0.5	-89
Turkey	19.3	6.5	12.8	3.2	0.3	0.1	0.0	0.6	-115
United Kingdom	17.3	12.3	4.9	4.7	0.3	0.0	0.0	0.2	2168
United States	18.3	12.7	5.5	4.7	0.9	0.0	0.0	0.4	156

Source: Arrow et al. (2002)

sources but are imported, they are not included as such in the genuine saving indicator. Of course, they appear in national accounts as imports but we have seen that environmental goods coming from poor countries are generally underpriced.

It gives a false impression that Belgium or the United Kingdom do not contribute to the depletion of the Earth's finite resources: resources that should be considered (from a moral point of view) the common patrimony of humanity.

Still, the most repugnant conclusion to be drawn from the figures is that the poor countries are overconsuming! Indeed, if they don't save enough, then they are naturally consuming too much. Actually, the genuine savings figures (here expressed as a percentage of their GDP) for most of the underdeveloped and 'in transition' countries are negative: e.g. Albania (−8.4 per cent), Armenia (−15.9), Azerbaijan (−24.4), Burundi (−7.3), Cambodia (−7.4), Chad (−7.4), Ethiopia (−11.3).

Arrow *et al.* are well aware of the problem:

> We [also] find evidence that many nations of the globe are failing to meet a sustainability criterion: their investments in human and manufactured capital are not sufficient to offset the depletion of natural capital. This investment problem seems most acute in some of the poorest countries of the world. . . .
>
> We would emphasize that insufficient investment by poor countries does not imply excessive consumption in the most important sense. For many of the poorest nations of the world, where productivity and real incomes are low, both consumption and investment are inadequate: current consumption does not yield a decent living standard for the present generation, and current investment does not assure a higher (or even the same) standard for future generations.
>
> (Arrow *et al.*, 2002: 26)

This is coming to acknowledge that not only composition but also scale matters, as Daly (1996a) argues, and that there can be too much consumption *and* investment as well as too little consumption *and* investment.

The shortcomings of the genuine saving index are such that one should refrain from drawing too definitive conclusions from these figures, but they seem to indicate that if the rich countries (*grosso modo* the OECD members) are probably saving enough to ensure no radical loss of welfare to the forthcoming generations, the fate of future generations in most Southern countries − including some of the 'emergent' ones − is likely to be even worse than the current generation's.

Conclusions

Sustainability and development are generally conceived as two different, yet related, matters. The former would deal with the biophysical possibility to generalize in time and space a given way of life while the latter would consist in asking the very different question of the desirability of doing so.

With respect to consumption and production patterns, and if one accepts the distinction between misconsumption as a critique of consumption from the consumer point of view, and overconsumption as another oriented criticism, misconsumption would be more of a development problem and overconsumption more of a sustainability problem.

Sustainable development is about both: what does our consumption entail for others, what does it do for us? Admittedly, to some of us it brings deception, discontent, meaninglessness, alienation, but it is also true that to some it brings pleasure, enjoyment, excitment and satisfaction. We do not all have the same unique conception of the good, valuable life. And we don't have to. But as human beings, we share moral intuitions. As Kant once said:

> Two things fill the mind with ever new and increasing admiration and awe, the oftener and the more steadily we reflect on them: the starry heavens above and the moral law within.
>
> (Kant, 1788)[19]

The 'moral law within' dictates that we care for others, for those who are not exposed to the risk of misconsumption as defined here, and those for whom our consumption levels and patterns may mean poverty, exploitation and the prospect of future shortages.

If the real sustainability problem is not to know if our way of life can be generalized but if it is compatible with a decent human life for others, now and later, then there is definitely something wrong with consumption.

Notes

1 'The greater part of our ills are of our own making, and [that] we might have avoided them nearly all by adhering to that simple, uniform and solitary manner of life which nature prescribed' (J.J. Rousseau, 1754, *What is the origin of inequality among men and is it authorized by natural law?*).

2 An allusion to Zaccaï's contribution to this volume.

3 Dotted arrows and text boxes are the author's additions.

4 According to Colin Campbell (1995), what is specific to modern consumerism is not the accumulation but the search for novelty, the desire to experience every novelty.

5 K. Marx (1867) *The Capital*, 1, Section 7: XXII. This sentence is much more expressive in the French translation: 'Il est vrai que le travailleur fait sa consommation individuelle pour sa propre satisfaction et non pour celle du capitaliste. Mais les bêtes de somme aussi aiment à manger, et qui a jamais prétendu que leur alimentation en soit moins l'affaire du fermier?'

6 Nowadays, the total expenditures in advertising in the United States amount to an annual bill of about 170 billion dollars (Bordwell, 2002).

7 Actually, the distinction between real and false needs can be traced back to J.J. Rousseau and his second discourse.

 8 This volume. He also discusses F. Hirsch's concept of positional good and its implication for 'social limits' to growth. This may be considered a fourth argument against the belief that more consumption means necessarily more welfare.
 9 These definitions differ from those proposed by T. Princen (2002).
10 A question equivalent to the one Durning (1993) asked: 'How much is enough?'
11 And perhaps also other species.
12 In this context, appropriation 'encompasses not only the act of asserting or establishing ownership of a resource but also the act of consuming resources without any proprietary claim' (Wasserman, 1998: 537).
13 Unfortunately, 'dematerialization' of Northern production and consumption patterns will not necessarily bring good news for them. Most probably, it will exercise an extra downward pressure on the prices of environmental goods.
14 On the various definitions of consumption, see Crocker and Linden, 'Introduction', in Crocker and Linden (1998).
15 What follows is largely inspired by Arrow *et al.* (2002).
16 One of the most challenging scientific problems for sustainable development is how far these different kinds of assets are substitutable.
17 Except for social capital.
18 This presupposes an adhesion to a weak sustainability standpoint, which accepts substitutions between different forms of 'capitals'.
19 E. Kant (1788) *Critique of Practical Reason* (T. K. Abbott translation) available online: http://philosophy.eserver.org/kant/critique-of-practical-reaso.txt.

3 Sustainable household consumption

Fact, future or fantasy?

Anton J. M. Schoot Uiterkamp

Introduction

Organisms in ecosystems are roughly divided into three groups with very distinctive functions: producers (e.g. plants), consumers (e.g. birds) and decomposers (e.g. dung beetles). In physiological terms producers perform anabolic functions, consumers metabolic functions and decomposers katabolic functions. In ecosystems terms human beings are involved in metabolism so they are consumers. Of course people produce goods like cars and take them apart again when they are no longer useful. Yet car producers essentially transform raw materials and energy and the same holds for car drivers and car scrappers. During the human production–use–recycle and waste-handling phases, environmental resources like metal, water, space and fuel maintain their quantity but change quality.

Besides environmental resources, human beings need each other to produce goods and services. Collective human metabolism, or social metabolism is not only subjected to biophysical and technological conditions but also to socio-psychological, institutional, demographic and historic conditions. The smallest socio-economic unit is the household with its corresponding household metabolism. In developed countries during the last century, household numbers have increased even more rapidly than the number of people. Moreover, the consumption of households has increased to an all-time high in developed countries and several developing countries are rapidly catching up. The ever expanding human societies and their households share the non-expanding planet Earth with all other living organisms.

Ten years ago these facts and observations induced us to ask the following questions: What is sustainable household consumption? Is past and present household consumption in the Netherlands and other European countries sustainable? And if not, what should be done to make it sustainable? We participated in three large-scale interdisciplinary national and international research programmes aimed at answering the questions: HOMES (Noorman and Uiterkamp, 1998), Greenhouse (Wilting *et al.*, 1999a, 1999b) and ToolSust (Kok *et al.*, 2003).

- *HOMES* (HOusehold Metabolism Effectively Sustainable) was designed to investigate household consumption patterns in the Netherlands in relation to environmental impacts, from the perspectives of environmental natural science, economics, policy science, psychology and spatial planning science.
- *The Greenhouse Project* was part of the Netherlands National Research Programme on Global Air Pollution and Climate Change (NRP). This multidisciplinary project aimed to analyse future options for reducing energy use and thereby greenhouse gas emissions by changes in household consumption patterns.
- *ToolSust* is a multidisciplinary project within the fifth EU Framework Programme dealing with sustainable cities. ToolSust is an acronym for 'The involvement of stakeholders to develop and implement tools for sustainable households in the city of tomorrow'. The project aimed to develop tools which focus on household consumption behaviour and which may be used to improve sustainability at the city level. Four European cities were selected to develop, test and implement these tools: Fredrikstad (Norway), Groningen (the Netherlands), Guildford (UK) and Stockholm (Sweden).

  In the following sections the answers to the three questions above will be given largely based on results of the HOMES, Greenhouse and ToolSust projects.

What is sustainable household consumption?

Before defining sustainable household consumption we should first define consumption, since consumption has different meanings in different disciplines (Stearns, 2001).

The natural scientists' meaning

As referred to above, matter or energy can neither be produced nor consumed. Consumption is therefore essentially transformation of matter or energy. The transformation is a 'downhill process' implying the appearance of pollution and a decrease in the utility of the transformed resource. In other words consumption gives rise to environmental impacts. These can be 'direct impacts' (e.g. CO_2 emissions from the combustion of coal for electricity generation), or 'indirect impacts'. For example, consider the consumption of bread. The pesticide emissions resulting from the production of wheat which is used for the baking of bread are indirect impacts of the consumption of bread. Similarly energy use can be *direct* (e.g. natural gas used in home heating), or *indirect* (e.g. diesel oil used by farmers is an indirect energy input in the production of cotton jeans).

The economists' meaning

Economists define consumption as part of total economic activity: it is the total spending on consumer goods and services. The rest of economic activity consists of capital goods investment.

The ecologists' meaning

To ecologists plants are producers, and humans and animals are consumers. For example, humans consume materials that are produced by green plants in photosynthesis.

The sociologists' meaning

Consumption is not precisely defined in a sociological sense. This is reflected in terms such as 'conspicuous consumption' and 'consumerism'. The latter term is formulated as 'a society in which many people formulated their goals in life partly through acquiring goods that they clearly do not need for subsistence or for traditional display' (Stearns, 2001).

The anthropologists' meaning

Anthropologists acknowledge the ambiguities associated with the very word 'consume', since it suggests both an enlargement through incorporation or intake and a withering away. Consuming is thus both enrichment and impoverishment. The latter is even more literally reflected in the wasting disease that used to be called 'consumption' and is nowadays better known as tuberculosis (Brewer and Porter, 1993). Douglas and Isherwood (1979) define consumption as a use of material possessions that is beyond commerce and free within the law.

Summarizing, various disciplines define consumption differently. Some definitions have a neutral connotation; others are more normative. Stern (1997) tries to combine the various meanings by presenting the following definition:

> Consumption consists of human and human-induced transformations of materials and energy. Consumption is environmentally important to the extent that it makes materials or energy less available for future use, moves a biophysical system toward a different state or, through its effects on those systems threatens human health, welfare or other things people value.

Taking the common definition of sustainability into account (WCED, 1987) we call a consumption 'sustainable', if consumption aimed at meeting the needs of future generations is not prevented by the consumption of current generations.

Is past and present household consumption sustainable?

We concluded from the results of the HOMES and Greenhouse projects that household consumption in the Netherlands during the fifty-year period since the Second World War was not sustainable. We also concluded that the future demand for resources due to household consumption will exceed the sustainable supply of environmental resources assuming the simultaneous need for an acceptable level of environmental quality. By the latter we mean that meeting human needs may not result in undue harm to other species and their ecosystems. The ToolSust project only confirmed the findings. The conclusions were drawn for a variety of levels and household functions. The geographical levels ranged from single households in the Netherlands to cities in four European countries. The actor levels ranged from political (taxes, regulation), economic (product quality, prices) and spatial (spatial planning and infrastructure) to the sociocultural level (institution and social norms). Household functions addressed ranged from fulfilling basic biological needs (e.g. feeding, shelter, transportation, entertainment, tourism) and fundamental social needs (e.g. substance, affection, understanding, participation, identity, freedom). Energy use was chosen as a proxy for the environmental impact of consumption although other indicators such as water use, waste production and habitat deterioration are useful parameters as well. Energy use manifests itself at three important environmental functions: sources, sinks and life support. Most energy is still derived from non-renewable fossil sources (e.g. coal, oil and natural gas). Energy use from fossil fuels gives rise to major environmental impacts such as smog, acidification and global warming. The latter problem results, among others, from overload of atmospheric sinks. Last but not least, global warming, acidification and eutrophication all deteriorate local, regional and global ecological life support systems which are necessary in providing functions such as oxygen production, water purification and plant growth.

Energy use related to household consumption was determined by means of Input–Output Energy Analysis (IO–EA) and hybrid analysis. IO–EA is a top-down method in which macro-economic monetary data on production categories are converted into energy data on final demand categories (e.g. consumers). In hybrid analysis data on production processes are used to determine energy requirements for all goods and services consumed by households. Our studies show that 70 to 80 per cent of national energy use is related to household consumption patterns. Direct household consumption energy requirements (e.g. natural gas, electricity and motor fuel) are substantial but they are often surpassed by the indirect household energy requirements attributed to the consumption of goods and services by households. The most important indirect energy use categories are food, transport and recreation.

Figure 3.1 shows the total (direct + indirect) average energy requirement in the year 2002 of households in four European cities: Groningen, Guildford, Stockholm and Fredrikstad (Kok *et al.*, 2003). The energy requirements range from 257 GJ for Groningen to 327 GJ for Guildford. The cities in countries with the highest expenditures – Norway and the UK – also have the highest

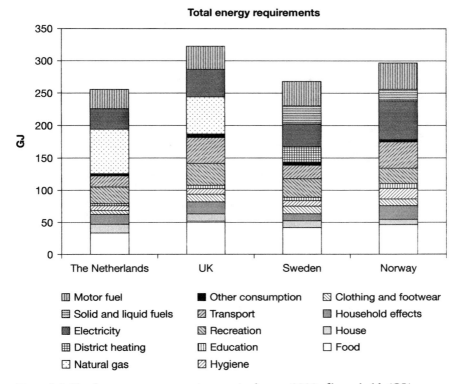

Figure 3.1 Total average energy requirement in the year 2002 of households (GJ/annum per category) for cities in the Netherlands (Groningen), United Kingdom (Guildford), Sweden (Stockholm) and Norway (Fredrikstad)

Source: Kok *et al.* (2003).

energy requirements. The share of indirect energy requirements is high in all countries, ranging from 49 per cent in the Netherlands to 60 per cent for Norway. Food, transport and recreation are important categories due both to high shares in the indirect energy requirement and the high energy intensities.

What should be done to make household consumption sustainable?

Before answering the final question, it is useful to review briefly a number of theoretical approaches regarding determinations of lifestyle, environmental behaviour and environmentally compatible production. Our starting point is the IHAT equation, a variant of the well-known population-based IPAT equation (Ehrlich and Holdren, 1971):

$$I(mpact) = H(ousehold\ number) \times A(ffluence) \times T(echnology)$$

According to this equation one may influence the overall environmental impact by either addressing the number of households (H), the consumption level per household (A) or the technological means employed in producing consumer goods and services.

The so-called Ecological Modernization Theory (EMT) (Spaargaren, 2000) starts from the assumption that environmental problems need to be addressed by the technological and organizational capabilities of modern societies. The EMT originated among environmental sociologists. They argue that environmental technologies are important in bringing about more sustainable means of industrial production and consumption. Several objections were raised concerning the EMT. It is not a classical 'theory' that can be rejected or accepted, and Glasbergen (2002) considers it a container concept that adds little analytical value to our set of environmental policy tools. In contrast, Scherer (2004) expands the EMT by proposing an interlinked network of the ecosphere, the technosphere and the sociosphere. In doing so she specifically addresses the need to include other species and ecosystems in designing sustainable technological and social systems of production and consumption, thereby giving environmental quality aspects their rightful place in the framework of sustainable consumption.

Environmental economists (Van den Bergh and Ferrer-i-Carbonell, 2000) argue that in studying consumer behaviour and seeking a comprehensive perspective on the limits and opportunities of sustainable consumption, at least five levels of 'consumer behaviour' need to be considered: preferences, motivations, constraints, decisions based on given preferences, goals and constraints (e.g. environmental tax policies change not only prices but also incomes which will subsequently affect behaviour: this aspect is often referred to as 'rebound', the unwanted change in behaviour which offsets technological efficiency improvements) and types of decisions relevant to environmental impact assessments (e.g. buying, use, reuse, recycling, repair, illegal dumping and waste treatment).

Environmental psychologists argue that behavioural change can be achieved by either employing psychological or structural strategies (Steg and Buijs, 2004). Psychological strategies are aimed at changing perceptions, beliefs, attitudes, values and norms. Giving feedback on the consequences of individual behaviour may be effective but it may be even more effective in the case of strong group behaviour or cohesion. Feedback combined with commitment (e.g. in so-called ecoteams) may overcome the voluntary nature of psychological strategies. Structural strategies are aimed at changing the context in which choices are made – examples are setting prices and making laws and regulations. Their effectiveness strongly depends on enforcement control and monitoring. Physical strategies may include offering energy-saving light bulbs, environmentally friendly products or providing access to 'green electricity'. Their effectiveness may be offset by rebound phenomena. Any type of intervention will be more effective, feasible and acceptable if it is systemically planned, executed and evaluated. An example is the 'DO IT' principle (Geller, 2002).

'DO IT' refers to the four steps in the intervention process: Define, Observe, Implement and Test.

Intercountry comparisons of households in the ToolSust project clearly showed the importance of industry, government and social institutions in determining the impact of consumption and thereby in suggesting possible energy-saving options. For example, the electricity supply system and the public transportation system are largely beyond household-level control.

Households are inclined to purchase energy-saving appliances, and to change food consumption patterns if the prices remain acceptable. However, they do not want to reduce their number of appliances, reduce mobility patterns or change holiday destinations. Relevant lessons for cities are that consumers think that energy and waste handling are more important than traffic while cities think the opposite. Moreover, cities often have logistic problems with providing organic food and consumers lack knowledge about labels. City administrators should improve both the communication within their organization and the external communication with their citizens.

The issue of trust deserves specific attention. Trust comes on foot and leaves on horseback and is often the only basis on which consumers act. Once their trust has been abused it is very difficult to restore it. Think of the example of global change in which scientists are often perceived to support a position pro or con depending on their place of employment. Consumers may react by completely disregarding the global change debate and its relevant outcomes.

Conclusions

We conclude that consumption practices are not only guided by deliberate choices but are largely driven by social constraints, routines, habits, household resources and facilities within households. Therefore tailor-made information aimed at specific households is preferred over general information or mass communication channels. Environmentally driven behavioural household change can be facilitated or stimulated by structural changes in institutional or geographical levels (e.g. city, national or international). Technological changes continue to play an important role as well as change in values. The rebound phenomenon is always just around the corner. Governments, societal institutions, special interest groups as well as industrial and service organizations are also important as are educational efforts.

The consequences for producers, consumers and policy-makers are clear. Transitions to (move) sustainable household consumption take dedication, compassion and patience from all parties concerned and for a prolonged period that is measured in decades rather than years. After all, the transition is not only taking place in Western countries but is occurring simultaneously throughout the world. China is a case in point (Liu *et al.*, 2005). The very fact that all countries share a single planet with specific environmental resources and rising populations with increasing expectations makes the drive to sustainable households all the more urgent.

Acknowledgements

I thank NWO, NRP and the EU for (partially) funding the programmes.

I also thank the members of the HOMES, GreenHouse and ToolSust project teams for their dedication, ingenuity and insight throughout the years. The contributions of Henk Moll, Rene Benders, Rixt Kok, Klaas Jan Noorman and Sanderine Nonhebel deserve special mention.

4 Sustainable consumption and sustainable welfare

John Lintott

Introduction: the ecological (and equity) argument against consumption

This chapter takes as its starting point the existence of ecological limits, and the need as a result for those in high-income countries to reduce their overall consumption. The argument that growth in production and consumption faces ecological limits goes back to Malthus and beyond, and in its modern form, to the 'limits to growth' debate which began in the early 1970s (Meadows *et al.*, 1972; Cole *et al.*, 1973). Essentially the argument is that exponential growth cannot be sustained in the face of ecological limits which are fixed, or which can only be pushed back to an extent which is itself limited.

There has been much debate, not only about how close we are to ecological limits, but also about how far it is possible in effect to push the limits back through improvements in the efficiency with which resources are used, thus continuing to increase output while at the same time reducing ecological impact. In other words, how far do ecological limits translate into limits on consumption?[1]

The extent to which ecological limits require consumption limits is largely the subject of books such as *Factor Four* (von Weizsäcker *et al.*, 1997), *Greening the North* (Sachs *et al.*, 1998) and *Natural Capitalism* (Hawken *et al.*, 1999). While these and other studies (such as Flavin and Lenssen, 1995; Jackson, 1996) have pointed to the enormous potential for improving the ecological efficiency of production, they have also tended to accept that such improvement by itself is not enough. Thus, for example, according to Weizsacker, one of the *Factor Four* authors:

> A factor of four improvement is considered to be readily achievable; a factor of ten is what may actually be necessary to achieve sustainable development.
> (reported in ENDS Report 272, September 1997)

Perhaps the key consideration is that as long as consumerism – implying ever-rising consumption – holds sway, no reduction in ecological impact per unit of consumption will solve the problem: on present trends any such reduction will be offset by greater consumption, mostly by the relatively rich.

It follows that it is not enough to improve the efficiency of production in order to achieve more consumption for less ecological damage; it is necessary to improve the efficiency of consumption so as to achieve more welfare for less consumption. And it is necessary to end consumerism, and not merely to reduce the ecological impact associated with a particular level or pattern of consumption.

These arguments apply to the world as a whole. In addition, few, even among critics of environmentalism, claim that rich country levels of consumption can be generalised to the whole world, except possibly on a timescale of centuries. One estimate (Daly, 1996a: 170–171) is that a thirty-sixfold increase in ecological efficiency (output for a given ecological impact) would be required for the world to achieve the same consumption per capita as the US without increasing ecological damage – even without further consumption growth in the US. Therefore, if consumption is to rise in the lowest income countries, so as to eliminate absolute poverty and reduce inequality, there is a need for much *stricter* limits to consumption (and perhaps large reductions) in high-income countries. If poverty cannot be abolished through the growth of output it can only be abolished through its redistribution.

There is thus a strong case, on ecological as well as equity grounds, for substantial consumption reduction among the relatively rich.[2] The persuasiveness of this case is undermined by uncertainty regarding where the ecological limits lie, the complexity of various ecological impacts, the potential for offsetting technological change, and the claims of rival experts. This makes it impossible to be precise about what a 'sustainable' level of consumption would be. However, such precision may be unnecessary: questioning rich country levels of consumption leads us to reconsider *all* the welfare impacts involved, including not only the ecological damage and risk, but also the costs in terms of labour, and especially the supposed benefits of consumption. It turns out that there are strong indications that, once basic needs are satisfied, further increases in consumption do not lead to improvements in welfare. Apparently, the welfare enjoyed in high-income countries can be obtained for a far lower level of consumption and ecological impact, one which would – hopefully – be sustainable, even if generalised across the whole world. A policy aimed at sustainable *welfare* may thus also imply sustainable *consumption*.

In a sense, what is required is a kind of cost–benefit analysis of different development possibilities, where the costs (ecological and other) and benefits (welfare) of the present, growth-centred approach are evaluated, and compared to alternative strategies (though I am not suggesting that this can or should be done in monetary terms, as in conventional economics). Ecologists sometimes seem to be suggesting that the ecological cost of consumer society is so enormous and so certain that we do not need to worry about the rest of the evaluation: we must reduce consumption irrespective of the welfare implications. Others question the scale of the ecological cost, and point to the uncertainties already mentioned. The case for reducing consumption, if it is to be widely persuasive, must rest on an evaluation of both costs and benefits, and of implications for both the ecosystem and welfare. The ecological and welfare arguments against consumption reinforce each other. After all, if ecological risks are small, then perhaps we should continue on the present path given that it is politically easier; while if ever greater consumption is essential to our well-being, then perhaps we should concentrate our efforts on mitigating the ecological costs without reducing consumption.

In what follows I shall consider first the evidence that consumption increases do not lead to overall improvement in welfare, and then discuss some of the reasons why, if this is so, we continue to seek increased consumption. Finally, I will attempt to draw a few provisional conclusions about the prospects for moving in the direction of sustainable welfare – and sustainable consumption.

The welfare argument against consumption

Consumption is not a goal in itself. As Schumacher put it,

> since consumption is merely a means to human well-being, the aim should be to obtain the maximum of well-being with the minimum of consumption.
>
> (1974: 47–48)

While there is no doubt that some consumption is essential to welfare, there is much evidence to suggest that any link between the two is the exception rather than the rule in rich countries: evidence from historical and cross-cultural comparisons, from studies of self-reported happiness, and from commonly accepted social indicators. If that is the case, then it is perhaps possible to reduce consumption while maintaining or even improving welfare, even leaving aside the ecological costs of consumption.[3]

Historical and cross-cultural comparisons

Orthodox economics encourages us to see the wish to consume as much as possible as almost part of human nature. Historical accounts of how consumer society came about and comparative studies of consumer and non-consumer societies suggest a very different picture: consumerism is historically specific. Sahlins' (1974) analysis of the 'original affluent society' documents the situation of Stone Age hunters and gatherers who, even in the marginal habitats to which they are nowadays restricted, require only a few hours a day to meet their needs, and thereafter stop work. This type of economy, while only possible at low population densities, suggests that, whatever its roots, consumerism does not stem from human nature. Moreover, there is much evidence of resistance to luxury in many traditional cultures (Stearns, 2001, ch.3).

The Industrial Revolution saw a transition from 'enough is enough' to 'the more the better', rooted in production for the market rather than for oneself.

Without mechanisms for expanding needs, people are inclined to work less rather than earn more, as demonstrated by the need for early industrialists to cut pay in order to force workers to work a 'full' working day (Gorz, 1989, ch.9). Or, as Sismondi expressed it in the early years of the Industrial Revolution, 'Luxury is not possible except when it is paid for by the labour of others' (quoted in Smith, 1993: 188); self-employed artisans chose leisure over luxury.

In fact, consumerism[4] only really became established during the post-Second World War period. Until then the finiteness of consumer wants – and therefore the likelihood that productivity growth would result increasingly in leisure rather than consumption increases – was largely taken for granted, and the Victorian and interwar periods saw the popularity, in many countries, of ideas about leisured utopia. At the same time there were fears on the part of the elite about what people would do with their free time. Increasingly there were other problems, such as selling an ever-growing volume of goods, largely unrelated to needs, and redistributing available work in the face of rising productivity and unemployment; meanwhile the 1930s slump undermined the desire for free time. The 'solution' adopted was based on engineering increasing demand for consumer goods. Working hours were to be somewhat shorter, though still far longer than they need be, thus providing both the time and income required for bouts of concentrated, commercialised leisure (Cross, 1993).

As a result a set of consumer values has emerged, which Lewis Mumford characterised as follows:

> There is only one efficient speed: faster; only one attractive destination: further away; only one desirable size: bigger; only one rational quantitative goal: more.
>
> (quoted in Sachs, 1992: 120)

Of course adopting these values is a recipe for frustration, since they require attaining not any particular state of affairs, but a continuously receding goal.

Subjective well-being

Subjective well-being and income: the evidence

Perhaps the simplest approach to investigating whether, within consumer society, well-being improves along with increased consumption levels, is to ask people. A large body of research on income and 'subjective well-being' (SWB) has accumulated over the past four decades, based on surveys of self-reported happiness. In these surveys, respondents are asked to grade themselves on a scale of happiness or satisfaction. For example, in the Eurobarometer Survey carried out in the EU since 1973, people are asked whether they are 'very satisfied, fairly satisfied, not very satisfied, or not at all satisfied' with their lives (Oswald, 1997: 1819).

A variety of possible factors explaining well-being have been investigated, in particular income. Although there is some variation in methods as well as in the detailed findings, there is broad consensus regarding income and well-being (one recent summary is Frey and Stutzer, 2002):

1   In a given country at a given time, there is some correlation between income and well-being: the rich are happier than the poor. A typical partial correlation coefficient however is only about 0.13 (Frank, 1999: 112), indicating that other factors are more influential. Factors such as health, social relationships and leisure activities seem to be particularly important (Argyle, 1999).

2   Comparisons between countries yield a more mixed conclusion. For low-income countries, there does seem to be some correlation – though again, rather low – between income and well-being; but there is no correlation among high-income countries. One typical finding is that:

> Above $13,000 in 1995 purchasing power parity there is no significant linkage between wealth and subjective well-being.
>
> (Inglehart and Klingemann, 2000: 171)

3   The most striking result however is that where income and SWB are tracked over time, there is no correlation whatever between the two. The proportion of US respondents who declare themselves 'very happy' has fluctuated between 30 to 40 per cent over the period 1972 to 1991. Japanese self-reported happiness was flat for nearly three decades (1961– 1987) in spite of more than quadrupled income (Frank, 1999: 72–73). In contrast, EU data (1973–1998) show large fluctuations (and big differences among member countries) but no clear trend and no relationship to income (Inglehart and Klingemann, 2000: 167).

The pattern shown by these results was first pointed out by Easterlin (1972), and dubbed the 'Easterlin paradox': happiness appears to be related to income on the basis of cross-sectional, but not time series, data. The explanation, suggested by Easterlin himself, and many others since, seems to be that it is principally income *relative to others* which determines happiness. Diener, for example, concludes that 'the influence of income is largely relative; it is not the absolute level of goods and services that a person can afford' (1984: 553).

This explanation in turn raises other questions, discussed below, in particular why relative income should be so important, but also what the relevant comparison group is.

It has also been suggested (for example, by Myers, 1992, ch.3) that income, relative not only to others but also to one's own previous experience, may have some importance for happiness. Equally there is strong evidence that pay rises (and indeed both 'good' and 'bad' events in general) have only a temporary effect. Moreover, fast-growing economies seem to show the same invariant happiness level as slow-growing ones, as the example of Japan, referred to above, illustrates. It is worth adding that many other, non-consumption, goods escape the logic of relative comparison (Frank, 1997): when individuals enjoy such things as more free time, shorter commuting, reduced stress, more extensive social networks and better environmental conditions, they generally experience greater well-being irrespective of others' experience. But these are things which often become scarcer as consumption increases.

The SWB approach: criticisms and limitations

Such studies of course raise methodological problems. SWB research is based on the idea that *reported* well-being is an accurate and unbiased measure of actual well-being. But this requires that there is, at all times, a particular level of well-being that individuals experience; that they are aware what that is; and that, faced with the survey situation, they simply decide to tell the truth. Yet we know little about the mental processes that may lead respondents to give particular answers. Some attempts have been made to establish a more reliable link between reported and actual well-being. There have been some findings that reported happiness is related to brain activity (Inglehart and Klingemann, 2000: 165). Reported well-being has also been correlated with various measures of physical health, as well as other reported feelings such as stress or self-confidence. Some care has also been taken to ensure that linguistic differences do not affect international comparisons (Layard, 2003).

However, research also suggests there are major weaknesses in the SWB approach. First, the concept of SWB itself is far from satisfactory. If measured at one point in time (as it usually is), it is extremely volatile. Survey respondents only form a judgement about their well-being once they are asked about it, and their response is highly dependent on how the question is framed, what other questions may have preceded it, and what past events they call to mind when answering (Kahneman, 1999; Schwartz and Strack, 1991). Second, respondents commonly deceive themselves and others by claiming they are happier than they really are (the 'social desirability' effect) (Hagedoorn, 1996; Phillips and Clancy, 1972); although, to complicate matters, such deception may in fact enhance actual well-being (Naess, 1994). Third, there may be a 'satisfaction treadmill' rather than a 'hedonic treadmill', where people evaluate their well-being, rather than their income, by comparison with others (and/or with previous experience) (Kahneman, 1999); in this case actual well-being may increase with income, without this being reflected in reported well-being.

It is thus likely that reported well-being partly reflects factors other than actual well-being. It is difficult, for example, to exclude this kind of possibility when considering the large differences among EU countries referred to above. Such problems, while serious, are unlikely to invalidate completely SWB comparisons over time in the same country, so that the key finding, that well-being does not increase with income, in the aggregate, over time, still seems persuasive. However, in the absence of a reliable relation between actual and reported well-being, it would be unwise to rely exclusively on the SWB findings.

Evidence from basic indicators

In view of doubts about the validity of the most direct approach – attempting to measure SWB itself – it is worth looking at other evidence regarding income and welfare. Here the obstacle is that there are no generally agreed objective measures of overall well-being. However, commonly used indicators of basic welfare (such as life expectancy, literacy, adequate diet) tend to stabilise at much lower levels of consumption than those of present-day Western Europe or North America. In some areas (family and community life, urban environment) welfare may decline as consumption rises.

A common pattern, illustrated by Figure 4.1 in the case of life expectancy, seems to be that for low levels of per capita gross domestic product (GDP) (and consumption) there is quite a strong relationship between consumption and welfare; past a certain point however there is little or no relation. But there are also many examples of low-income countries that nevertheless achieve high welfare levels. On the basis of these data a per capita GDP of about purchasing power parity (PPP) $5,000 (the world average is over PPP$7,000) would be more than enough to ensure a similar level of welfare as even the most overdeveloped countries manage. And perhaps a far lower per capita GDP level would be sufficient where other circumstances are conducive to well-being

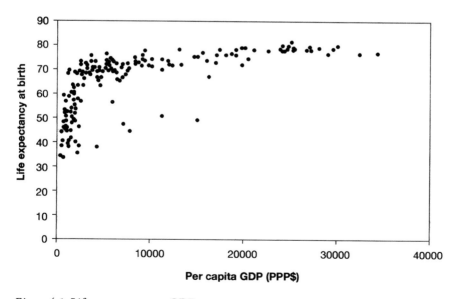

Figure 4.1 Life expectancy vs. GDP

Source: Based on UNDP 2003. Data are for 2001.

(including in particular relatively equal distribution of the GDP). A similar pattern exists for other basic indicators, such as literacy, per capita daily calorie intake, and access to safe drinking water.

Of course such indicators of basic needs satisfaction cannot be taken to indicate well-being as a whole; they merely reflect a consensus about its most basic aspects. The problem is that as one moves from the most basic aspects of well-being, conceptual and measurement problems increase, and there is little agreement about what well-being measures should include. That said, indicators such as suicide rates, crime rates, marital breakdown rates and rates of mental illness do not suggest that rising income and consumption are associated with improved well-being.

What the evidence about consumption and welfare suggests

Evidence about the relation between consumption and welfare is thus not absolutely conclusive, and perhaps can never be, given the general and intuitive nature of concepts like welfare and well-being. Nevertheless, all the evidence points in the same direction, and taken together it is very persuasive. The goal of ever-increasing consumption, far from being universal, is mostly a feature of post-1945 industrialised societies. Even where consumerism rules, factors other than consumption are more important for welfare and – most important for our argument – insofar as welfare is related to consumption at all, once basic material needs are satisfied, it is an individual's relative, not absolute, consumption level that counts for his or her welfare. It follows that in rich countries increases in consumption do not and cannot, in the aggregate, lead to improvements in overall welfare.

If increased consumption does not lead to greater welfare, the search for welfare on the other hand seems to lead to consumption: if, for whatever reason, individuals are concerned about their consumption relative to others, then a (quite rational) individual wish for greater welfare will drive a process of ever-increasing consumption, even though it cannot lead to any increase in welfare in the aggregate. Each individual increases his or her own consumption so as to improve his or her relative position; but the aggregate result of each individual's efforts is to maintain his or her relative position, and thus welfare, on average unchanged.

But there is one kind of evidence that appears to contradict the rest, and which finds expression in the economic doctrine of 'consumer sovereignty': most consumers apparently freely choose, at considerable sacrifice in work and effort, to consume as much as opportunities allow and certainly more than is required to meet their basic needs. This suggests that perhaps, in spite of the evidence presented, consumers may experience welfare as a result of rising consumption. If the view that consumption does not contribute to welfare is to be defended, and also if we are to find acceptable mechanisms to reduce consumption, we must answer the question: If consumption does not bring welfare, why do we consume?

Why do we consume?

Consumerism has become central to modern society, and has insinuated itself into every area of life. Mechanisms for promoting consumption are many and mutually reinforcing. Values, institutions, socialisation, all, at least in part, tend to promote the consuming individual. Even growth itself engenders growth: according to Baudrillard it is the 'negative goods, nuisances compensated . . . which play the role of economic engine' (Baudrillard, [1970] 1998: 41–42), thus completing the circle, in which economic activity creates nuisances, which in turn stimulate further economic activity. One may perhaps even speak of an 'ecology of consumerism'. Thus an exhaustive answer to the question why we consume would entail examining how we grow up to be consumers, as well as the role of marketing and the media, and much else.

A more limited goal is to explain why relative consumption is so important for welfare. Some explanations focus on relative *status*. Some authors (e.g. Frank, 1999, ch.9) have suggested an evolutionary explanation: concern with status pays off in evolutionary terms, because higher status individuals are more likely to survive and reproduce. Even if this is true, however, it does not explain why status competition should be expressed in the form of ever-increasing consumption. As we have seen, consumerism is not a feature of most societies. On the other hand, sociologists (e.g. Bourdieu, [1979] 1984) and anthropologists (e.g. Douglas and Isherwood, 1979) have tended to focus on the role of consumption in forming and expressing identity, and in marking status and group membership. While no doubt such factors contribute to an explanation of how pressures to consume operate *in a consumerist society*, they do not explain why the search for status and identity should be channelled into ever-increasing consumption in the first place. As one author suggests, there are contrasting sources of identity in traditional and postmodern societies: in the first case, 'history, tradition, community, ancestors and extended families', and in the second, shopping and consumer products (Sardar, 1998: 138).

Instead of considering these approaches further, I want to focus on another explanation, Hirsch's theory of positional goods, which explains the importance of relative consumption for individual welfare in terms of individuals' attempts to satisfy quite ordinary, almost universal, wants. This is not to underestimate the importance of socialisation, marketing or status competition, but none of them would perhaps be as effective if they were not anchored in ordinary self-interest.

Hirsch and positional goods[5]

Fred Hirsch, in his discussion of the 'social limits to growth', focuses on the increasing importance of 'positional' goods as societies become richer (Hirsch, 1977). His theory is an attempt to give an 'economist's answer' to certain questions (although it has not had much impact on mainstream economics). In particular, he asks:

Why has economic advance become and remained so compelling a goal to all of us as individuals, even though it yields disappointing fruits when most, if not all of us, achieve it?

(Hirsch, 1977: 1)

Hirsch is also 'economic' in terms of assumptions: his theory does not require that consumers be greedy, envious, obsessed with status, or in any way irrational (although of course they may be). On the contrary, the theory assumes that individuals are rational pursuers of ordinary goals such as good jobs, decent housing conditions and adequate transport, and that, as in conventional economic theory, they know their own interests best. The problem is with the feasibility of attaining such goals, and the ability of market and voting mechanisms to respond to these interests adequately.

Positional goods are goods such as cars, higher education and holiday homes in the country, whose contribution to each person's welfare diminishes as others acquire them. Such goods are subject to 'social congestion': the welfare they yield is 'restricted not only by physical limitations of producing more of them but also by absorptive limits on their use' (Hirsch, 1977: 3). Positional goods have grown in importance but individual attempts to increase the welfare derived from them result in no welfare increase overall. An increase in total welfare derived from positional goods is by definition impossible.

The force of this argument is widely recognised in the case of cars, for example. Increasing car use initially provides greater mobility and access, as it leads to the appropriate infrastructure of roads, garages and other facilities being put into place. But beyond a certain point it merely leads to congestion, and mobility is reduced (Sachs, 1992). Meanwhile the collapse of alternative means of transport, as well as the relocation of homes, workplaces and essential services further away from each other, drastically reduces access for non-motorists, and locks motorists into car use. Attempts to restore mobility by 'improving' infrastructure merely aggravate the problem. But, although the result may well be lower speeds (Illich, 1974) and certainly less access than prevailed before the car, at huge cost environmentally and economically, the incentive to drive remains: indeed, cars may become essential for most people.

A similar process is at work with other positional goods. Yesterday's suburbs become today's inner cities, and a Master's degree may be worth the same to a job seeker as a school-leaving certificate a generation or two ago. But as long as the original 'good' – a suburban environment, more qualified employment – is still sought, the incentive remains to move yet further out of the city, or to study even longer.

Hirsch's argument is thus not that individuals' preferences are irrational, but that the expression of individual preferences, however rational, may lead to an irrational outcome. 'Consumers, taken together, get a product they did not order' (Hirsch, 1977: 6). There is a 'tyranny of small decisions' (1977: 40). Furthermore, the perverse outcome stems, not from status-seeking or other motivations, but from the characteristics of positional goods themselves. Nor is it merely a result of market allocation as such: positional competition in fact tends to lead to pressure for collective provision (e.g. state education or road building), which merely makes the problem more acute.

How much consumption is positional? An accurate overall assessment of the positional economy would involve enormous, perhaps insuperable, conceptual and measurement problems:

> It has not been found possible to estimate over what proportion of economic activity social limits to growth are in play.
>
> (Hirsch, 1977: 181)

However, as the examples already given make clear, the positional argument applies with particular force to most aspects of transport (e.g. commuting and tourism) and land use (especially suburbanisation). These are particularly relevant for the environmental argument since they are important sources of ecological damage. In particular, transport is now the main energy-using sector in the rich countries, and increasingly so.

The strength of Hirsch's argument lies in the way it shows how individuals may be virtually forced into positional competition – irrespective of factors such as greed, envy, 'keeping up with the Joneses' or, on the other hand, frugality or ecological concern – simply in pursuit of ordinary goals such as access to basic facilities, a decent job or a cleaner environment. Hirsch describes a mechanism where concern with relative consumption drives welfare-less (but ecologically damaging) output growth.

Underlying the positional problem, Hirsch argues, is scarcity in a much stronger sense than that which applies to other goods. The latter are scarce in the relative and temporary sense that other satisfactions must be given up for them and thus supply is fixed at any given time; but more can be produced tomorrow if demand is sufficient. Positional goods, on the other hand, are associated with absolute scarcities.

Common experience suggests there are many situations where mechanisms similar to the positional one are at work, where each person strives to consume either more or the same as others. There are situations similar to positional competition, where overconsumption occurs: military arms races, over-harvesting, environmental pollution (Frank, 1999, ch.10). This is accentuated by a variety of mechanisms, or 'social traps' (Cross and Guyer, 1980), where factors such as short-sightedness and ignorance play a major role, facilitating patterns of behaviour which entail immediate and obvious rewards, but longer term, more uncertain, costs. This is the case with addictive consumption, and with consumption which involves getting into large and unnecessary debts.

Consumption and community

However, more generally, there are many other ways in which others' consumption affects one's own, and may thus fuel concern with relative consumption, and therefore drive growth. Indeed, the combined effect of various factors – among them the way tastes are formed, and the fact that much consuming is done in the company of others – means that supposedly individual preferences are inherently social. Thus, for example, the individual experiences hunger; but social influences in effect determine what he or she expects to eat. The result may then be a society where everyone eats meat, without being better off than in a vegetarian society. The omnipresence of the social element in consumption gives strong support to the argument that it would be possible – as a society – to consume less without loss of welfare.

If consumption choices are largely determined by social influences, then the persistence or otherwise of consumer habits may be associated with the nature of community. Wachtel (1983), for example, argues that in modern societies the desire for more arises from the attempt to compensate for loss of security derived from community and tradition. He emphasises the collective nature of consumerism as well as anti-consumerism:

> the concrete realities of our society . . . make it difficult for all but the most extraordinary individual to extricate himself from the temptations and exigencies of the consumer life on his own. From street crime to a shortage of public recreational facilities to peer influences on oneself and one's children there are a range of forces that make individualistic and consumerist choices hard to eschew. Understanding how our present choices are self-defeating is a crucial step in the process of change, but so too is understanding how the social and political context makes such self-defeating choices seem almost inevitable. . . . One of the reasons that for a short time in the Sixties young people could so radically alter their lives is that they did so jointly, with mutual support and within informal social structures that made it easy.
>
> (Wachtel, 1983: xiii)

This raises the question of what the relevant comparison group is when assessing the well-being derived from a given level of consumption. With the development and internationalisation of communications and media images, individuals are less likely to be satisfied with the same standard of living as their neighbours, and more likely to demand, or at least dream about, a Hollywood lifestyle. On the other hand, the role of consumption in expressing identity, group membership and status does not mean there is a need to consume lavishly, and social pressures may be used *against* consumerism. The development of low consumption communities, where individuals gain status by consuming less rather than more, as with the 'voluntary simplicity' movement (Dominguez and Robin, 1992), may be an essential part of any attempt to reduce consumption without sacrificing welfare.

Conclusions

The starting point of this chapter is that ecological limits, together with global equity, require that high-income countries reduce their level of consumption. The available evidence suggests that welfare, in the aggregate, does not improve as consumption increases. Thus a solution to the problem of attaining a sustainable level of consumption is in principle available and desirable: large cuts in consumption are possible, resulting in reduced ecological damage, without loss of welfare. Moreover the transition to sustainable consumption – and welfare – may entail other welfare gains which have not been discussed here, such as reductions in employment time, and perhaps, as a result, increased conviviality and closer communities. The transition may also entail substantial, though temporary, costs of adaptation, both psychological and material.

On the other hand, the evidence also suggests that consumers are acting quite rationally, as individuals, in seeking increases in consumption. In these circumstances it may be largely futile to ask them, as individuals, to consume less. Collective decisions are required that change the environment within which individuals in turn make choices. Unfortunately the ballot-box may do no better than the market: individuals who, in their role as consumers, drive cars and seek better educational qualifications, are likely, in their role as voters, to vote for more roads and more schools.

Equally there is little prospect of using economic incentives in order to achieve sustainability, the cornerstone of orthodox environmental economics, if sustainability requires limiting or reducing consumption (overall, and not only in specific categories). There is an obvious contradiction involved in limiting consumption by encouraging certain types of behaviour with the prospect of higher income, and therefore more consumption. A post-consumerist society would be one where economic incentives no longer work.

However, in view of the present dominance of the profit motive, these considerations could easily lead to a pessimistic conclusion – that nothing much can be done to tackle consumerism and achieve sustainable consumption. On the other hand, environmental pressures will continue and grow, imposing increasing costs, monetary (defensive expenditures) and in terms of quality of life. Meanwhile there is much evidence from experimental economics that, even in an individualistic society, people's first instinct is to cooperate rather than compete. Summarising this evidence, Gintis suggests that:

> economic actors are not self-regarding, but rather in many circumstances are strong reciprocators who come to strategic interactions with a propensity to cooperate, respond to cooperative behaviour by maintaining or increasing cooperation. . . . H. Reciprocans is thus neither the selfless altruist of utopian theory, nor the selfish hedonist of neoclassical economics.

> Rather, H. Reciprocans is a conditional cooperator whose penchant for reciprocity can be elicited under circumstances in which personal self-interest would dictate otherwise.
>
> (Gintis, 2000: 311–316)[6]

Perhaps then there is a potential willingness to cooperate to reduce consumption, rather than compete by consuming more, if ways can be found to mobilise it.

Of course, the end of consumerism requires a major shift in values and institutions. But meanwhile, there is at least one broad area – transport and land use – where existing pressures may lead to change, and which illustrates the point that the environmental consequences of consumption can become sufficiently serious to overwhelm the various pressures to consume – at least in some cases. There is also one consequence of reducing consumption – shorter working hours – which may make it more immediately attractive, in spite of positional pressures. And there is one policy area where change is absolutely necessary if a policy of reducing consumption is to be acceptable: redistribution of income and wealth.

Transport and land use

In attempting to reduce consumption, it makes sense to begin with areas where consumption is largely positional, as well as particularly harmful ecologically. Transport systems have a particular importance in view of their large and increasing contribution to energy use and greenhouse gas emissions, as well as their largely positional nature. Moreover the costs of car use in many urban areas, in terms of congestion and commuting time, have become so great and so obvious that support for change may be relatively easy to obtain.

The general principles of a sustainable transport policy are not very controversial, and include some goals which even existing policy supports in principle. Most fundamentally, there is much scope for reducing the need for people and goods to travel at all, notably by a refusal to allow further suburbanisation, and through a shift towards production for local use, and away from global trading. The first of these is official policy in the UK and elsewhere, but applied in a half-hearted way; the second however runs directly counter to government policy, which favours increasing trade in most countries.

Where there is a need for transport there is great scope for reducing its environmental impact, by moving away from car use in cities and for commuting, and towards collective transport, cycling and walking; and by moving away from road and air travel, and towards trains and water transport for long-distance transport of both people and goods. Government policy in most countries is, again, ambiguous: in principle favouring public transport in cities, but unwilling to curb long-distance travel by car or air.

In terms of future environmental pressures, the problem is that while increasing congestion will continue to force change in and around cities, there is less immediate pressure to discourage car use elsewhere, or to slow down the increased long-distance transport of goods and people.[7]

Shorter working hours

Reducing consumption implies, other things being equal, reducing aggregate employment hours. Although there is of course great concern when this takes the form of unemployment, reductions in the length of the working day, week, year and life have almost universally been regarded as progress, progress moreover that, unlike consumption increases, does not necessarily imply an ecological cost. A strategy of reducing consumption provides an opportunity for this progress, which has stalled in recent years, to resume. A very substantial reduction in employment time, perhaps eventually by half or more, can be envisaged, with a multitude of resulting benefits in terms of well-being (Hayden, 1999; Lintott, 2004). Moreover, cutting working hours along with income may provide one of the more feasible and promising routes to lower consumption, since any costs of adjusting to lower consumption may be offset by the benefits of freeing up time for other uses.

There have of course been reductions in the working week since the nineteenth century, though they have slowed down since the 1960s. There have also been other reductions in lifelong working time (e.g. increases in holidays and in time spent in full-time education, lowering the retirement age), and in the 1990s there has in some cases been a return to the previous trend of reductions in the working week, notably in France with the introduction of a thirty-five-hour week. But this type of policy has often been constrained by conflict over maintaining pay level. If it is part of a strategy of reducing consumption, however, the whole point is to reduce pay along with hours. What is needed is a legislative framework which allows workers to claim work time reductions, such as the right to job share (with corresponding pay cuts), or to unpaid leave, from employers when they wish; this would allow workers, as a first step and if they chose, to enjoy the benefits of increased productivity in the form of reduced work rather than increased pay.

Redistributing income and wealth

Economic growth has provided a pretext for ignoring problems of inequality: in the conventional view poverty may be tackled more effectively through growth than by redistribution, and this is of course politically convenient. If however we are to move towards limiting, and eventually substantially reducing, average consumption, redistribution of income and wealth becomes essential – a process similar to the 'contraction and convergence' advocated in the case of carbon emissions. If absolute poverty is not to be abolished through growth, then it can only be abolished through redistribution. Moreover, relative poverty, by definition, can only be reduced though redistribution. Reducing inequality implies a reversal of existing trends in most countries, and even existing policy in countries like the UK which have reduced taxes on higher incomes over the past twenty-five years. But without such a reversal no policy to reduce average consumption is acceptable.

Notes

1   'Consumption' is used here in the conventional economics sense, to refer to the purchase of goods and services. It is of course possible to use the term differently, for example, to mean consuming *resources*, or, on the other hand, consuming the *services* which purchased goods provide. The economics definition is more useful however if one is to be able to distinguish between the efficiency of production – the ratio of consumption to ecological cost, a question of technology – and the efficiency of consumption – the ratio of welfare to consumption, a question of lifestyle.

2   This case is made in more detail in Carr-Hill and Lintott (2002, ch. 2).

3   There is of course a great deal of debate about what such terms as welfare and well-being mean. For the purposes of this chapter I avoid this debate, and generally accept the conventional economics view that individuals are the best judges of their own welfare: welfare (used here interchangeably with well-being, satisfaction and happiness) is whatever individuals want. Furthermore it should be the objective of policy to deliver welfare in this sense, and to do so efficiently: the most welfare for the lowest ecological and other costs. However, economists also usually claim that individuals' wants are adequately expressed as preferences in markets (and in some cases as votes in elections), and this claim is rejected in what follows.

4   In its modern sense, implying ever-increasing consumption (and ecological impact). Consumerism, in the more restricted sense of spending on luxuries or display, may be found at various times and places, but on a limited scale, and restrained by religion and tradition (Stearns, 2001, ch.2). The concern here is with not only the superfluity but also the insatiability associated with modern consumerism.

5   This section draws on earlier discussion in Lintott (1998).

6   Gintis also finds evidence of '*loss aversion*' and '*status quo bias*', which bears on the argument for a major reduction in consumption habits, though in an ambiguous way: 'individuals often prefer the status quo over any of the alternatives, but if one of the alternatives becomes the status quo, that too is preferred to any of the alternatives' (2000: 315).

7   Certainly the removal of perverse price incentives (although it is unlikely to solve the problems by itself) seems desirable. This applies to cars, where currently vehicle tax and insurance premiums are generally payable per time unit rather than per distance travelled, thus drastically reducing the marginal cost of individual trips to the driver. It applies even more dramatically to air travel, where fuel is currently exempt from taxes.

5   How to attribute power to consumers?

When epistemology and politics converge

Grégoire Wallenborn

What consumers?

Sustainable development policies assume that consumers can and will evolve towards more sustainable consumption patterns. An implicit hypothesis of changing consumption patterns is that consumers have the power to change their behaviours and lifestyles, or at least that they have relative power to do so. In the framework of sustainable development, a lot of institutions are endowed with specific power and special capacities: political institutions, law institutions (property, legislation, rules and procedures), economic institutions (monetary, finance, accounting) and techno-scientific institutions. The power devoted to consumers is however much less clear, for this power depends on a disciplinary perspective: different powers and capacities coexist and sometimes oppose each other. The definition of the consumer is political by nature, since it corresponds to different perceptions of society or of what society should be. For instance, the debate between efficiency and sufficiency reflects points of views that distribute powers differently among institutions and individuals.

Consumers are often attributed with the power to become informed, to assimilate the information they obtain and to translate it into acts. But are we not giving them too many powers? Are we not confusing *citizens'* powers with *consumers'* powers? To what theoretical model does this attribution of powers refer? The power attributed to persons depends on the definition being used, because their attributes result from such definition, and there are numerous definitions of consumption which give rise to different models of consumers. The different ways of viewing consumers give rise to as many different ways of defining them and attributing to them different capacities for action. The power attributed to consumers is therefore linked to the theoretical framework within which one works and from which research hypotheses are drawn.

To give an example, the definition of the 'green consumer' is not an easy task and depends on the theoretical view that emphasises or neglects some relations. Is a consumer green by his or her attitudes or by his or her actions? Somebody who buys ecological products can do it for other reasons than the protection of the environment (for health or money, for instance). To determine whether a consumer is ecological or not, his or her freedom to achieve an action can be a good test (Pedersen, 1999). These actions can belong to different consumption sectors (mobility, food, energy) or reveal an effort the consumer is ready to make (waste sorting, will to pay more for some products for their environmental quality). If one of these options is taken, it will implicitly tell how sustainable consumption is defined. However, what is freedom in the consumption realm? Numerous constraints weigh on consumption activity: incomes, knowledge, social norms and cultural codes. These constraints explain why consumers often feel powerless when faced with environmental problems.

We might also recall that the idea of 'the consumer' is an abstraction, because the consumer is also (generally) worker, spectator, citizen and so on. All these facets form parts of the same person and breaking up an individual into these different aspects is, in itself, a way of distributing powers, as I will suggest throughout this contribution. Reisch (2001) shows the different temporalities linked to the fact of being either consumer, or citizen, or worker. There is also a need to discuss not only the interrelations between consumer, citizen, person, individual and human being, but the topic of consumption itself, which combines material, knowledge, spiritual, psychic and cultural dimensions.

The consumer (noun) is sometimes referred to as 'responsible', 'civic-minded' or 'committed' (adjectives). This suggests that a state of fact can be corrected by describing it morally and giving it a strong personal morality. The philosophical and political question is how this 'moralisation' can be extended beyond some exemplary cases, and above all, *what principle* gives this morality any power whatsoever. Moralisation is a way to attribute power to consumers. Critiques of the consumption society are also made in assuming power to individuals in one way or another (Wilk, 2001).

At a very simple level, the power of action of an individual is defined by either external relations or internal relations. The attribution of capacities or power is the usual working of scientific theories. Based on Latour (1989) and Stengers (1993), one may say that the way a science constructs its object of study is a way of giving it *power*, a certain capacity to respond to the precise questions researchers ask. While it is obvious that the researcher's theoretical framework is crucial in determining the type of questions raised, I would also like to demonstrate that the power attributed to the object (in this case the consumer) depends upon this theoretical framework.

The way a scientific discipline defines its object of study is performative in nature; defining the consumer in a given way is tantamount to obliging him or her to behave in a given way. In other words, theories are not politically neutral. A theory is a process of qualification of certain actors, and at the same time of disqualification of others – if only because they are neglected. The definition of acts of consumption implicitly reveals their relation to acts of production: any definition of consumers also implicitly refers to a way of attributing power to the producer, because the consumer's definition indicates the producer's hold over the consumer. Inversely, Dobré (2002; see also Chapter 11, this volume) shows how the definition of consumers is important in trying to zero in on their 'power of resistance'.

In the following sections, I will analyse the different ways of defining consumers and the powers that may consequently be attributed to them. A scientific discipline is a way to exhibit certain relations and to neglect others. In this perspective, an attribute is a fundamental relationship that gives sense to the considered approach. Different inventories of theories about consumers have been made (Wilk, 2001; Shove, 2003): this contribution aims at analysing powers attributed to consumers. Five kinds of disciplines will be analysed according to the relations attributed to the consumers: ecosystems, market, personality (internal relations), situation (external relations) and infrastructures (humans are peripheral). Vocabulary is the marker of disciplines. The following markers will be used respectively: natural resources; preferences, rational choice; attitude, behaviour; representation, practice; structures, objects. From this short list, one can see that these approaches sometimes overlap; different dimensions may be present in a given theory (as, for example, in the social psychology). It is however important to notice that power distribution differs in each approach. The approaches described hereafter are obviously sketchy, and should be taken as a construction of contrasted points of view. McGuire (1999) has argued in favour of an epistemological perspectivism: each theory captures a part of the reality, though it misrepresents it. In order to avoid relativism, I will add a political dimension to the analysis.

Consumers as living beings

From a biophysical point of view, all living beings consume natural resources, whether raw materials and energy or other living beings taken in the ecosystem. From this standpoint, the contemporary problem of consumption by humans is its excessively high level, with too great a material flow. In thermodynamic terms, overconsumption is an excessive flow of matter and energy. This point of view is developed in particular by Princen (1999) in order to analyse the conceptual roots of consumption.

This approach has the merit of rooting human activity in ecosystems, and as such remains a crucial horizon of studies on sustainable consumption patterns, at least for setting material limits to it. It nevertheless fails to take into consideration the specific characteristics of the human species and notably omits all non-biophysical reasons why humans consume. Consumption is perceived as a 'transformation of matter and energy', and households are the units that operate this transformation of flows. Accordingly, this organic and materialistic vision of consumption reduces consumers to 'black boxes' through which pass flows that they deteriorate. From a biophysical point of view, consuming first and foremost means to destroy, whereas from a sociological point of view consuming means to create ties and identities.

This approach does not give much power to consumers, other than that of deteriorating ecosystems. It none the less has the merit of integrating the environment into the socio-technology system, thus making it possible to set out new conditions for needs validity. The debate on needs – determining which are 'real', to be distinguished from 'false needs' – has been around for a long time and has always taken a moral posture, but the conditions of the truth (and falsehood) of needs have changed. Today, even if needs are considered 'real' from the individual consumer's point of view, they can be 'false' from the collective consumer's point of view. The biophysical approach reveals the power of the 'collective consumer', namely the human species. The gauge of truth thus evolves as humanity extends its hold over the planetary environment.

Consumers as rational economic agents

Attributing power to someone is initially a fiction, but this fiction can become a reality if it is reflected in behaviours suited to the fiction. Certain definitions have a performative effect: they transform the object to which they attribute a power or a capacity for action. To some extent, this is the case with economics. Consumers are defined as rational by economic theories and in fact are invited to be increasingly rational if they wish to benefit from the 'best rates' available on a given market. For instance, services markets are characterised by the great variety of products in a given sector: comparing the available products requires considerable capacities for researching information and equally considerable capacities for analysing it. Consider, for instance, insurance company or mobile telephone company rates. The performative effect can sometimes be ironic: in the hope of saving 10 or 20 euros, 'rational consumers' can be obliged to waste hours of their time.

The radical statement that individuals are the basis of our contemporary society is partly linked to the increase in consumption. Individuals are increasingly independent of other individuals: relations to others are more and more abstract (formalised and institutionalised), while relations to objects are growing (Røpke, 1999). Consumption is thus deepening the process of consumption itself through the individuation of markets and objects. Increasingly segmented markets characterise the dynamic of consumption in our society today. This segmentation obliges consumers to learn new 'signs' in order to identify the products that interest them (the very rationale of labels). But this segmentation results from different models of the consumer which, by trying to tap emerging consumption trends, define and oblige consumers to position themselves (even temporarily) in one of these segments.

Based on these examples, we may observe that the attribution of cognitive capacities (rationality or faculty of recognising signs in a profusion of objects) to persons also renders the market complex. Markets are more and more a universe of signs, and that is possible only if consumers are assumed to be able to treat all this information, to decode all these signs.

In the classical economic approach, the consumer is something of a rational automaton (referred to as *Homo economicus*): fully informed, maximising its interests, with fixed and stable preferences, he or she makes calculated choices that are revealed through purchases and awareness of prices. The economy explains how individuals spend their financial resources, how they evaluate different possibilities and how they take purchasing decisions with the purpose of maximising their satisfaction. According to the theory of rational choice, the individual supposedly knows all his or her needs and how to satisfy them. His or her decision is supposed to be independent of the different institutions (except the market) and to consist in a maximisation of utility within budget constraints. The consumer is sovereign in his or her preferences, and his or her choices are expressed in the elasticity of prices.

The strength of this model resides in its performative effects, including the selection of policy instruments. Since the consumer is rational, if his or her purchases do not correspond to theoretical projections, it is because he or she is poorly informed. Thus, it is crucial to give adequate information to consumers. Many implemented policy instruments are oriented towards information, since one believes that consumers' choice is based on information.

In the economic model, the main attributes given to consumers are instrumental rationality and insatiable needs for goods and services (Stagl and O'Hara, 2001). Power is however mainly devoted to economic analysis in its perceived ability to make forecasts and especially to define democracy as individual freedom. Indeed, according to this approach, nobody can tell individuals what they have to do. It is obviously a highly theoretical view, and furthermore prevents the establishment of political common aims (Norton *et al.*, 1998). The flaws of this approach are well known and have often been criticised (e.g. Siebenhüner, 2000; Niva and Timonen, 2001). This model fails to take into account a number of constraints or relations: the role of institutions and of social relations, the different temporalities and spaces (that explain diversity of preferences and their sometimes fast evolution). For instance, the assumption of self-interested behaviour is contradicted by the fact that people help each other altruistically. Social and environmental norms are also respected even though people do not benefit from them. The economic model of the consumer is thus static, failing to explain changes of preferences and of consumption patterns; it is individualistic, failing to explain why people make a commitment to collective actions; it is utilitarian, failing to take 'non-consumption' choices into consideration. The *Homo economicus* model credits the consumer with a unique and narrow rationality which is focused on the act of buying. This act is only one instant in the process of consumption. In general, however, at least three separate roles exist in this process: the influencer, the purchaser and the user (or consumer). These three roles may but are not necessarily played by the same person. To be succinct, we could say that the definition of the influencer comes within the realm of psychology while that of the consumer comes within the realm of anthropology. These two approaches extend the notion of rationality, each with its own methods. Rationalities are plural, and psychology shows it through internal relations, while sociology explores different situations and coherent value systems. The power of consumers is therefore made relative by placing them in situations in various contexts: psychological, social, institutional, historical, philosophical and so on. In the following sections I will examine how consumers are given internal and external attributes.

Consumers prompted by their attitudes

The influencers are the actors who influence buying behaviour. These 'actors' can be of different types: family and friends, advertising, conscious or unconscious motivations. The psychological analysis will attempt to identify the various *motivations* that underpin a given behaviour. As Jackson and Michaelis put it:

> practitioners in the [economic] field have been considerably more inquisitive about the nature and origins of consumer motivations than economic theorists. New areas of inquiry such as consumer psychology, marketing and 'motivation research' have developed a rather rich body of knowledge – a 'science of desire' (Dichter, 1964) – for producers, retailers, marketers and advertisers wanting to know how to design and sell products that consumers will buy.
>
> (Jackson and Michaelis, 2003: 23)

Numerous studies try to explain environmentally responsible behaviour through the analysis of attitudes (see Rousseau & Bontinckx, Chapter 6, this volume).

Stern (1992) has developed a very comprehensive model of behaviour relative to the environment. He describes a hierarchy of eight levels of causality. The highest level is that of the most general causality: background factors that include everything related to the social situation and social relations. Level seven concerns structural and institutional factors, namely external conditions, incentives and constraints (rules, technology, convenience and rewards). Level six is related to fundamental values. Level five designates the general attitude towards the environment. Level four concerns attitudes to and standards for relevant behaviour in favour of the environment. Level three describes knowledge and beliefs specific to the same behaviour. Level two refers to the commitment to adopt such behaviour. Finally, level one designates the behaviour itself. Two things stand out in this hierarchy of levels of causality:

1   attitudes determine behaviour, but attitudes have more sway over actions if those actions are easy and not costly;
2   the individual is at the heart of the analysis, and the natural and social environment are peripheral.

The main attribute given to consumers is their attitudes, namely a relatively permanent manner in which an individual thinks or behaves when confronted with a given problem. Attitudes are set at a deeper level than opinion, less rational, and exhibit a certain endurance. Attitudes then are acquired but difficult to change.

Psychological models give more than their due to attitudes resulting either from rational intentions or from unconscious processes (of which personality is a central factor) (Pellemans, 1999). A personality is defined as an intra-attitudinal structure. If a specific attitude contradicts the behaviour related to it, one speaks of 'cognitive dissonance', because the subject is assumed to be coherent. Another important aspect of psychological studies is the formation of attitudes: whether through experience or through discourse, the acquisition of an attitude is fundamental for linking it firmly to a behaviour. Brown and Cameron (2000) suggest that to change values influencing consumption patterns, consumers' attitudes should be changed first. The definition of attitude, however, implies that consumers' attitudes evolve slowly.

The study of attitudes allows one to construct scales such as the New Environmental Paradigm. Attitudes cannot be observed directly. They are known through three types of responses: cognitive, emotional and behavioural. Most of the time, the measurement of attitudes is based on statements made by the individual questioned (Sabourin and Lamarche, 2005). Such statements depend in large measure, however, on the context in which they are made – whether in an individual interview, a group discussion or a poll – and the person conducting the survey defines the context in which the responses are expected. The survey creates a specific context for statements: the person surveyed is forced to express him or herself in another's words and the resources he or she can mobilise fall more within the 'citizen's' category. Thus attitudes and behaviours do not stem from the same context and may lead to considerable incoherence.

Of course, different survey methodologies seek to limit such bias, so that the invitation to speak will result in the genuine expression of the attitude or capacity of action of the person surveyed. The fact remains, however, that when consumers are defined through their attitudes and behaviours, the idea is to try to determine their underlying *motivations*. This causal model is useful for explaining one-off acts such as purchases. It is considerably less productive, however, for more complex behaviours.

The way psychology constructs the consumer is related to the fact that this discipline is fundamentally a laboratory science. Psychological devices try to expose some relations between determined variables. As is well known, the problem of a laboratory science is to extend facts and phenomena beyond the walls of the laboratory. A way to bypass this problem is to recreate the context of the laboratory (Latour, 1989). To some extent, the shop is this context where consumers' behaviours are controlled and experimented with, even though it seems that marketing strategies are more empirical than based on laboratory results. Marketing is indeed always more powerful in contexts where the link between motivation, attitude and behaviour is simple, namely shops. Everything in a shop acts as an incentive to make this connection! Insofar as the purchaser is not as rational and well informed as the economic model assumes, marketing uses this 'deficit' to make the consumer's power respond directly to the prescription.

The psychological approach defines a consumer's power through his or her individual choices – whose rationality is more or less extended – and by his or her dispositional capacities. Therefore it includes the economic model. When behaviours and attitudes are specific, a model may be developed. When behaviours and attitudes are general, however, they have to be included in a wider context that sociology or anthropology can tackle. In this case, relation to the situation is a better explanation of the observed behaviour. Some experiments in psychology show that dispositional capacities are more willingly attributed than situational capacities (Jones and Harris, 1967). Thus, ironically, psychology explains why individual determinants are often overemphasised, and why policy instruments are geared towards individual behaviours and attitudes.

Situations and practices

While psychology is based on laboratory work, sociology is grounded on field study. This means that different attributes and concepts are given to the objects of study. Attitudes and behaviours are variables denoting control and stability; representations and practices refer to contexts open to interpretation, since their relations are not a priori determined. It is interesting to note that questionnaires on a given subject are not organised in the same way by psychologists and sociologists. Usually the psychologist asks questions initially about attitudes and only then asks questions about behaviours, though the sociologist asks questions about practices and then questions about representations. Each would argue that he or she proceeds in this way to prevent biases. The former would say that it is easier to deceive about his or her attitudes than about his or her behaviour, and therefore attitudes have to be established first in order to prevent 'contamination' by previous questions. The latter would suggest that practices are more important than representations, and should consequently be collected as 'pure' as possible.

Defining consumers by their *practices* obliges us to study the contexts of the activities, routines and habits that shape the acts of consumption. Speaking about consumers' practices implies reproducing the social situations in which standards are an important constraint, as well as all the other actors (Boltanski and Thévenot, 1991). Standards are of different types and all deserve to be taken into consideration. Environmentally friendly actions acquire their meaning in a context whose rules are shared by others. The consumer's power to change consumption patterns depends on the situation, i.e. on the constraints that permit the reproduction of social practices. Practices refer to more ongoing actions than behaviours.

The concept of practice also offers degrees of freedom to consumers through 'narratives': consumers build their identity by articulating the beliefs and values they attach to their acts. For the anthropologist, exchange of goods and services organises the group and the diverse identities. Identities are formed through negotiation between oneself and others, and this is particularly at stake in products bought and consumed. This is also what is covered by the sociologist's concept of 'lifestyle': consumers bring their different social practices into a more or less coherent whole, particularly through the way they describe them. The logic of actions can vary widely, depending on the society, culture, fashions and situations. The sociologist therefore accepts the heterogeneity and paradoxes of behaviour more readily than the psychologist.

Beck and Giddens describe the contemporary society in attributing reflexivity to systems and subjects. Power is delegated to norms as well as to the capacity to reflect these norms by individuals and institutions. Individuals construct their identity through reflection of their practices. Giddens' (1984) theory of structuration suggests a connection between constraining structures and the competences of agents because the latter draw their dynamism from the structure's enabling dimension in the form of cognitive and normative resources and rules. Here, the capacity of agents to act intentionally (*agency*) by relying on available resources, to bring them into play in interactions, invites us to conceive of action as an activity with its uncertainties and its potential for innovation rather than as a practice always already adjusted to a binding context. Structuration is an ongoing and reflexive process of reproduction and trans-formation of social practices and relations in space and in time through the duality of structure. 'Structure is always both constraining and enabling' (Giddens, 1984: 25): it refers jointly to the concepts of constraint and capacity. Integrating a sociology of action, the theory of structuration defines social actors as competent. Competence is understood to mean all the things the actors know (or believe), tacitly or discursively, about the circumstances of their own action and that of others, and which they use in the production and reproduction of the action. This competence underlines in particular a reflexive capacity of human actors, continuously involved in the flow of day-to-day conduct, i.e. that they have the 'capacity to understand what they do while they do it' (Giddens, 1984: xxii). Yet this 'reflexivity operates only on partly a discursive level' (Giddens, 1984: xxii) and, in human competence, Giddens makes a distinction between discursive consciousness and practical consciousness. Discursive consciousness refers to everything 'actors are able to say, or to give verbal expression to', namely everything to which the concept of consciousness is frequently reduced. Practical consciousness, a more original concept, refers to 'all the things actors know tacitly about how to go on in the contexts of social life without being able to give them direct discursive expression' (1984: xxiii) and is not unrelated to the concept of routine. The border between these two modalities of competence is fluctuating and changing.

The interest of Giddens' theory is the elaboration of attributes both enabling and limiting, and at the same time the description of a dynamics that can make attributes evolving. Giddens thus believes that awareness of environmental problems may progressively shape individual lifestyles. From the viewpoint of a connection between practice and narrative, the consumer can find a new identity as a *citizen* and thus may seek to reconcile his or her values and practices in some situations. This viewpoint is exploited by 'ecological modernisation', developed in particular by Spaargaren (1997), an approach which gives an important role to environmental technologies and makes them more efficient.

The theory does not reside in the a priorism that the more sustainable organisation of production and consumption is incompatible with capitalist organisation. The role of individuals is important but forms part of a wider system: they do not act as isolated units but as co-actors. This approach leads to a reassessment of the organisation of production and consumption from a consumer-oriented standpoint. The concept of consumer society is therefore no longer considered a starting point for a critique of overconsumption, but on the contrary is recognised as a key concept for better understanding the dynamic of industrial societies. At the level of governance, it is recognised that society can no longer be regulated uniformly from a central point but that there is a need to build horizontal forms of policy development in which the target groups take an active part by choosing and developing their own 'path' to sustainability. In conclusion, this approach is rooted in the idea of sustainable development: without calling into question the structure of production–consumption, it develops a set of tools, among which participation and eco-efficiency are the cornerstones.

Ages and objects

So far I have outlined some of the main approaches that define 'consumers' and attribute certain powers to them. I would like to complete this review by presenting some theories that define consumers from other perspectives, namely relating to the socio-technology system or the study of objects. Objects and technology are indeed increasingly present and mediate relations between humans. Each age may be characterised by a set of techniques and objects; this infrastructure puts constraints on human capacities in triggering some relations and in restraining others. Traditional relations are now replaced by relations with objects (Røpke, 1999). It is then important to grasp the strong tendency which invites us to invest our wages in objects, and which has so many adverse effects on the environment.

The different sociological analyses of the 'consumer society' generally take as their starting point the observation that objects have more than a material status: they also have symbolic and immaterial dimensions. Consuming is a cultural act: we consume to set ourselves apart or to belong to a group; we consume signs and images (of brands) as much as products and uses. With the proliferation of objects in our society, they become mediators of social relations: it is *through* objects that people enter into relations with one another. The object of consumption is two-faced: on one side it refers to a function, a certain use value; on the other it is a sign that allows distinction, membership of a social group. It is not always easy to sort out the aspects of signs of distinction and value.

If we insist on the logic of signs and symbols, we are engaged in a *differentialist* model: the consumer buys and uses goods to establish his or her universe of standards and values as distinct from those of other social groups (Veblen, 1899; Baudrillard, 1970; Bourdieu, 1979; Douglas and Isherwood, 1979). Criticism of consumer society is thus never far off: objects are no longer linked to a function or a defined need and the act of consumption is described as an alienation – a criticism Marx tied in to object fetishism. This demonstrates in any case the importance of *values*, since to consume is also to consume signs.

Along with signs of distinction or social value, objects increasingly incorporate immaterial qualities and information that require new faculties on the part of consumers. We can think here of the different signs that indicate some aspects of the processes from which the product results (e.g. traceability, different labels) as well as the guarantees sold together with the material product. This dimension explains among others things the ever-growing product differentiation.

Many theories focus on the historical evolution of societies. Lifestyles and consumption patterns have changed profoundly in less than a century, in parallel with the entire socio-technology system. One of the most striking elements of this evolution is mobility (in particular cars) and the spatial planning that this implies. Numerous studies in the sociology of technology have demonstrated the extent to which society and technology mutually shape one another (e.g. Hughes, 1987; Callon, 1995). Accordingly, a very important variable in the framework of sustainable consumption is this flexibility. Geels and Kemp (2000) have developed the theory of transition management to determine the gradual move from one technology system to another. In such a perspective, the idea is to seize opportunities to make the socio-technology system branch off into a more sustainable direction. The technological system is probably the most important variable, but also the hardest to 'manipulate', given that individual consumers have very little control over the evolution of the system. If we consider that behaviours and practices are determined in large part by the socio-technology system, then the consumer does not have much power: his or her behaviours depend basically on the age in which he or she lives.

Since objects underpin questions on 'sustainable consumption', it is interesting to move away from humans and place objects at the centre of the analysis. The difference between production as work and consumption as satisfaction of a need seems to be rooted in our anthropological structure (at least in modern times). But consumption is also production in the sense of an appropriation and domestication of products. Over and above the definition of consumer, it is therefore important to consider the definitions of the object of consumption. Do not objects, in their materiality, also have the power to define consuming acts, and therefore also consumers? What do objects of consumption require from consumers to be appropriated? Criticisms of the economic model discussed thus far have focused on the description of the human world, but this criticism may be extended to the world of objects. In the socio-technical perspective, they are not as passive and indifferent as other models would suggest.

Conclusions

One way among others of developing a typology of the different studies on consumers is to consider the different types of variability of consumption choices. The choice of consumption depends of course on many factors, but these may be grouped according to (1) consumers' *personality* and *motivations*, their attitudes and behaviours, (2) consumers' social situation, where they live, social standards, territorial constraints, and (3) the age and society in which consumers live, the socio-technology system and lifestyles. This distinction (individual, space, time) can help us to account for the different constraint systems that shape consumption. This typology is inspired by Shove (2003), and Jackson and Michaelis (2003). Table 5.1 summarises the characteristics of four ways of defining consumers and the main consequences concerning sustainable consumption.

The psychological approach is more in line with the state of the market and of the financial and economic development that goes with it: the idea is to prompt consumers' attitudes and behaviours to evolve towards a supply that would become more diverse to offer ecological products. This approach also fits in with the current marketing trend. But a well-understood sustainable consumption policy is above all a policy of demand for products that are more respectful of the environment and of socially acceptable production conditions: the sociological approach – and in particular that of ecological modernisation – allows us to envisage the development of more eco-efficient technologies and their appropriation by consumers who are mindful of the stakes of sustainable

Table 5.1 Four ways of defining consumers and their consequences for sustainable consumption

	Automaton	*Personalities*	*Situations*	*Ages*
Theories and disciplines	Economy	Psychology Psycho-sociology	Sociology Anthropology	Sciences-technology-society
Consumer attributes	Rational Seeks to maximise his/her profit	Motivations Behaviours Attitudes Desires (conscious and unconscious)	Practices Narratives More or less coherent identities	Lifestyle determined relative to objects
Market	Supply and demand adaptation	Acting on supply	Acting on demand	Making the market branch off
Instruments	Information Prices	Information (marketing) Images Education	Regulations (including social norms) Empowerment of associations	Infrastructure Planning Strict norms

development. By comparison, economic and infrastructure-based approaches touch upon elements firmly rooted in the evolution of society and are as such harder to change, even if for that very reason they are even more crucial in nature.

According to the five approaches described above, consumers are acting through what defines them: the fact of being alive; computation; attitudes; social relations; socio-technical constraints. Consumers are accordingly diversely determined: their preferences are fixed, their attitudes are stable, their being is determined by social relations and technological infrastructure. In a way, their freedom is the capacity to escape from these determinations. Freedom is the term that permits to explain the unpredictability from the point of view of the various models or theories. In this perspective, freedom of the consumer could be defined as the ability to produce his or her own subjectivity in negotiating the different relations. Freedom is not the subject of this chapter however, but rather the power attributed to consumers. Power enables consumers to act upon others or upon objects: power is a hierarchical and external relation.

Considering the current societal trends, namely deepening individualism and profusion of objects, a consumer who would want to commit him or herself to sustainable consumption has first to resist the different determinations that define him or her today; he or she has to construct his or her attitudes and preferences in a radical, exogenous manner. At the individual level, consumers appear to have relative power, through consumption choices (including non-consumption) or through appropriation and domestication of objects. At the system level however, consumers' behaviours appear to be widely shaped by social and technological structures. There is here what could be called the paradox of collective action. The individual power of the consumer gets its meaning only if other individual powers are mobilised. The individual power remains virtual as far as it is isolated. This set of powers could be actualised throughout socialisation processes or devices of collective statements. Today, politicians and associations address consumers as individuals: how can they be prompted to switch to other, more collective identities? Apart from policy instruments focused on production, different pathways to 'socialisation' are not being sufficiently explored or used: direct and overall socialisation through norms laid down by politics; socialisation through the empowerment of civil society associations; the socialisation of consumers through objects of consumption themselves.

Markets are currently shaped by offer. How would it be possible to build devices that reveal the social demand for consumption? Isolated individuals are dependent on the offer. Social devices should then redistribute the relations between powers, for example, through non-governmental organisations (NGOs) or collectively organised persons. These devices would replay the relationships of the different facets of individuals: consumer, citizen and worker. In this perspective, moralisation would not come from outside (from a theoretical definition of individuals) but from the creation of values that are not pre-existing. Above all, the collective devices should be performative if ever they have an existence.

Part II

Who is sensitive to sustainable consumption, and why?

6   Testing propositions towards sustainable consumption among consumers

Catherine Rousseau and Christian Bontinckx

A change in production and consumption patterns is needed to launch sustainable development. In what way can consumption evolve towards this objective, and what impulses are required to stimulate the change? Is it realistic to hope that consumption will shift due to a voluntary modification of consumer choices? And if so, what measures are to be taken to encourage consumers to change? Or should one develop policies that can influence consumption patterns without active participation and impetus from consumers?

Environmental behaviour explained by motivations

If we consider that consumers can be, partially, a driving force for change to sustainable development objectives, we start with the implicit hypothesis that consumers have the will and the power to adapt their consumption choices and more generally their lifestyle. In Western societies, the quantity of products on the market is growing continuously, and so are the possibilities, incentives and pressure to consume. The market in which consumers must make their daily decisions is becoming more and more complex. The classic economical approach, based on a consumer who makes rational decisions in a stable, well-defined system of preferences, did not explain observed behaviours. It has progressively been abandoned for more dynamic and complex models driven by sociological (e.g. age, family situation, profession, lifestyle, socio-economic status), psychological (e.g. personality, motivation, perception, learning, attitudes) and situation-bound variables (e.g. the circumstances in which the buying decisions have to be made).

The socio-psychological approaches tend to explain consumer behaviour by motivation. By 'motivation' we understand all factors that generate a particular behaviour, and explain its direction, its intensity and its persistence (Moisander, 1999). In most of the studies that try to understand environmentally friendly behaviour, environmentally friendly choices are studied as an option adopted by people to contribute to a better quality environment. In this case the principal motivation of environmental behaviour is 'to protect the environment'. Moreover, research into ecological behaviour has often been conducted without carefully defining what exactly is meant by ecological behaviour. This is often presented as an undifferentiated class of behaviours (Poortinga *et al.*, 2004). By doing this, the different types of environmental behaviours are considered implicitly dependent on the same elements. It is interesting to look at the classification of environmental behaviours as proposed by Stern (2000). He suggests distinguishing the so-called environmental behaviours by either their intention or their impact on the environment. Classifying them by their intention means that the behaviour is defined by the actor's motivation to protect the environment, without considering its real impact on the environment. Certain behaviours may be adopted with the intent of reducing environmental impact, without necessarily producing a minor impact on the environment. The classification based on the impact does not focus on the actors' motivations but defines a type of behaviour by its environmental impacts. Gatersleben *et al.* (2002) have shown that behaviour adopted with the intention of protecting the environment is determined more by attitude-related variables, while behaviours with an environmental impact, such as energy consumption, depend more on socio-demographic criteria like the size of the family and income, which influence individuals' capacity to adopt a specific behaviour.

Authors such as Thorgersen and Moisander have underlined the role of personal factors like perceptions, attitudes and emotions in motivations to adopt a behaviour that contributes to a better quality environment. They considered, for instance, the perception of a moral responsibility towards the protection of the environment, the perceived normative pressure, the perception of the identity of a responsible consumer with regard to the protection of the environment and perceived behavioural control. Among these motivations, Moisander (1999) distinguishes primary motivations (that engage in a general behavioural class such as 'to adopt ecological behaviour') and selective motivations (that generate a particular behaviour: e.g. recycling, energy saving). She observes that the moral responsibility that people feel with regard to the protection of the environment and the perception of their identity as an ecological consumer are more powerful motivations than the others for adopting an environmentally friendly behaviour.

When studying consumers' attitudes and behaviour towards environmentally friendly products (products with an ecological label), Thorgersen (2000) considers that everyone who chooses an ecolabelled product has to pass through a sequence of mental stages: determining a personal objective with regard to environmental protection, believing that making responsible purchases is an efficient strategy to achieve this objective, being familiar with ecolabels (that they exist, what they look like, what they mean), having faith in the label; then, in the shop, paying attention to the labels and deciding to buy products with an ecolabel. The attention paid to ecolabels is influenced by the availability of ecolabelled products in shops and by certain elements that have to do with personality, i.e. the consumer's perception of their capacity to influence the achievement of the objective (in this case a better quality environment). The consumer's faith in the power of their consumption choices as an environmental strategy is influenced, among other things, by their attitudes and certain features of their personality.

The complexity in making a more ecological choice increases with the potential conflict that may occur between the different motivations which play a role in a specific consumer choice. Indeed, the 'ecologically responsible' consumer aims at two different goals in his or her consumption choices: on the one hand individual goals, and on the other hand collective goals and the protection of the environment in the long run. His or her choices imply two types of evaluation, and in the end prove more complex than the choices based on just one type of goal. The choices are all the more complex because conflicts between motivations may be associated with a 'social dilemma' (Moisander, 1999): the social advantage for the individual consumer who does not adopt the behaviour may be larger than the advantage he or she gains by adopting cooperative behaviour, independently of what the other society members do. However, all the individuals in the society profit less when everyone drops out instead of joining in. Even a 'green' consumer may be tempted to act as a 'free rider' because an ecological product costs more or takes more time, or because he or she thinks that his or her individual impact is only marginal.

We wanted to better understand how this conflict of motivations intervenes in consumption choices. We chose an approach 'impact', studying behaviours that have less impact on the environment and not the behaviours adopted in order to better protect the environment. We analysed the motivations of a variety of different individuals when they adopt this kind of behaviour. The results presented below come from a research project called 'Criteria and impulsions for change toward a sustainable development: sectoral approach'.[1] Among other things, this project aimed at a better understanding of the logic of attitudes and behaviour developed by consumers with regard to consumption choices that are compatible with sustainable development. In particular, we would like to verify the existence of variations in that logic for different sectors of consumption and to determine the sectors for which more possibilities of change exist.

Perceptions and motivations for sustainable consumption

Qualitative approach

We applied a qualitative psychoscopic method based on the realization of discussion groups and in-depth interviews. The procedure for data collection uses an open, no directive, permissive and indirect approach (Gauthy-Sinéchal and Vandercammen, 2005). This type of approach can offer a fine-tuned understanding of the variables that influence individual behaviour, such as beliefs, opinions, attitudes, motivations and aspirations.

In a primary phase we organized four discussion groups[2] to explore the following questions: How are the different dimensions of sustainable development (environmental, economic and social) perceived in the field of consumption? What opinions, perceptions and attitudes are expressed with regard to suggestions of a consumption that would be compatible with sustainable development? In which sectors of consumption is there a potential for the development of behaviours compatible with sustainable development? What changes in consumption behaviour are considered possible?

The studied concept was sustainable consumption and not only environmentally sustainable consumption. We also took into account the social dimension of sustainable consumption (e.g. fairness, Fair Trade, and respect for International Labour Organization (ILO) conventions in the production and distribution stages).

Participants were selected in order to obtain a maximum number of different logics.[3] The selection was made on the basis of socio-demographic criteria (age, gender, family size, children or no children, professional occupation, education) and one recruitment question about personal interest in environmental protection.

At the end of this first phase we selected five product categories responding to different qualitative criteria: environmental and social impacts of products, existence of alternative propositions in terms of products or behaviours, and favourable attitudes towards change among the participants of the focus groups. Those five categories were: washing detergents, electric appliances for textile care, electric kitchen appliances, jeans, and paint for indoor decoration. The participants showed interest in other themes such as food, energy and transport, but we did not make an in-depth study of these themes because they were included in other studies in the same scientific programme.[4]

In the second phase, four discussion groups[5] were organized to go deeper into the opinions, attitudes and behaviours for each of the product categories under consideration. This enabled us to determine two sectors[6] (textile care and indoor paints) for which a larger variety of consumers, in terms of socio-demographic criteria but also in terms of motivation profiles, might adopt more ecological behaviour. Next, two supplementary discussion groups explored different scenarios of change in those sectors. These scenarios were developed by the scientific project team and/or resulted from earlier group discussions. Thus engaging scenarios could be determined.

Finally, individual in-depth interviews were held with sixteen individuals selected on the basis of socio-demographic criteria, as well as on the basis of emotional expectations, to go deeper into the different behaviour logics.[7] The objective was to establish the obstacles and adoption probabilities for each scenario and to identify the scenarios that are most likely to be adopted by consumers belonging to a variety of attitudes and behaviour logics.

Psychological motivations for sustainable consumption

In all the groups, the concept of sustainable development could not be described or was unknown. None of the participants spontaneously gave a 'definition' that linked the environmental, social and economic dimensions together. After a short presentation of the concept, the participants gave their personal views:

they found the idea of sustainable development utopian or even incoherent (development versus sustainability) and vague. They tended to associate it with a variety of environmental problems and, to a lesser degree, with the social dimension. In the field of consumption, they considered these problems from the standpoint of their repercussions at different levels: the public that is affected (individually or collectively), the spatial aspect (nearby, far away or on a planetary level) and in time (present or future problem). The participants feel powerless to integrate the various dimensions (environmental, social/ethical and economic) of sustainable development in their consumption choices. They can only take into account the aspects of these dimensions that correspond to the dynamics of their personal motivations. In this perspective, the environmental dimension seems to hold more potential than the social dimension. In the field of socio-economic ethics (social dimension), the fear of proximity effects (such as higher prices of goods, delocalization of production and consequently of jobs) play an inhibiting role, while the perceived benefits of ethical consumption choices (such as respect for labour agreements, fair prices paid to producers) remain symbolic: the participants who make these choices hope, but are entirely uncertain, that their behaviour will have positive social implications. Neither the protection of the environment nor altruistic concerns seem to be sufficiently strong motivations, capable of competing with the other motivations that intervene in the consumption choices which the participants make. In most cases, the environmental or ethical criteria fall under a complementary 'option' that the consumer adds to the final choice: they are neither determinant nor decisive.

In general, participants find that it is not up to them, through their individual choices, to assume responsibility for environmental protection but that this is up to the public authorities and producers, identified as the main body responsible for the environmental deterioration. Participants who adopt environmental behaviours say that their most important contribution is to sort their waste. They think that responsibility for the products lies with the producers. The product characteristics, which, in their opinion, put more pressure on the environment, fall under production process. They systematically underestimate the environmental impact of their consumption choices and everyday habits. They express high expectations towards the public authorities and to a lesser extent towards the market: only products that respect the environment should be allowed on to the market, so that they are able to make choices according to their own criteria with the guarantee that they do not affect the environment, no matter what products they choose.

Nevertheless, when asked what changes they would make in their consumption choices, the participants had plenty of proposals. They gave many different possibilities for choosing (or not choosing) certain products, making more rational use of natural resources such as water and electricity, managing their waste, travelling and so on.

In fact, sustainable consumption does not appear to resemble an application of a single logic in the different fields of consumption, but rather a multitude of diverse options, varying with the individual logics and the sector in question. For instance, a participant who is concerned about status and image expresses his 'need' for distinction by purchasing stylish clothes; he does not buy 'ethical' clothes because they do not satisfy him in terms of distinction. But he expresses his altruistic sensitivity by purchasing other products such as food from Fair Trade organizations.

In the different groups we observed that the possible choices, in a multitude of environment-respecting behaviours, are evoked according to personal motivations. These motivations vary from individual to individual, and for each individual they can vary from one sector of consumption to another. Moreover, these motivations very often do not concern environmental protection, but fall under other motivations such as concern for the health of close relations or the need to belong to a group.

The reference framework that we used for interpreting the discussion group themes is the CirCept method and the Jungian psychic model.[8] Without going into details, the Jungian approach to the psychological types gives an understanding of consumers' motivations. The psychological types taken into consideration are mainly introversion and extraversion, with the understanding that these psychological types constitute the expression of a preference: introversion and extraversion are the extremes of an axis in which different compromises and tendencies exist.

In consumption matters, introversion corresponds with a profile that aims to satisfy internal needs by looking to the external world for things that could match them exactly. Extraversion aims to use the external world in general for the stimulation of internal sensations. When we classify and organize the themes that the participants evoked with regard to their perception of sustainable consumption, we could define three essential axes: the first is the time relation (now–distant future), the second concerns space (nearby–far away), and the third is about the affected population (me–all living things). By combining these axes with the introversion-extraversion psychological types, we obtain the matrix shown in Figure 6.1.

This matrix illustrates the tendencies that influence the elaboration of an attitude towards sustainable consumption. In fact, the spatial approach (e.g. the perception of the nearby or distant environment) is linked closely with taking account of the individual or of a small or larger group. Likewise, we can link the extravert attitude to a perception that gives priority to the present or to the near future, and the introvert attitude to a time perception that puts more emphasis on foresight and the future in the longer run. The analysis of the dynamics of the participants' attitudes and behaviours shows that attitudes and behaviours favourable to sustainable consumption develop on the basis of psychological motivations. Among others we observe three major motivations: control and concern about savings, protection of health and security, and the feeling of belonging to a group (normality, altruism, good citizenship). Three other motivations are expressed with less intensity: convenience, interest in new technologies, and the need to be distinctive.

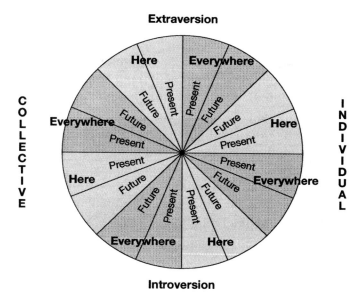

Figure 6.1 Matrix of the tendencies that influence the elaboration of an attitude towards sustainable consumption

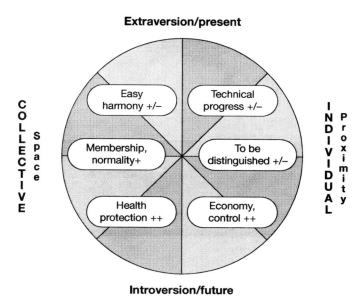

Figure 6.2 Comparative matrix of personal motivations with regard to ecological consumption

The different tendencies develop motivations of varying intensity for the adoption of sustainable consumption behaviour. In the matrix in Figure 6.2, the '+'–signs indicate the intensity of the relation between personal motivations that promote the adoption of choices respectful of sustainable development and the inclination to adopt these choices. Even though all the tendencies can develop environmentally friendly behaviour, we observe that introvert-like participants are much more motivated to make an effort than extravert respondents.[9] Let us take an example to show in what way different individual logics intervene in the adoption of ecological behaviours: one participant, who is very concerned with questions of ethical consumption as well as with environmental problems, purchases food products with a label. Nevertheless he cannot explain precisely what the meaning of these labels is or what effect this purchase has on the problems that preoccupy him. Another participant, who is also concerned about environmental problems, gives no credibility to labels and does not buy labelled products. His concern for the environment manifests itself more clearly in his energy consumption. In that field, his electricity meter is the only trustworthy reference. The first respondent is stimulated by seeking a sensation, an option that is coherent with his perception of sustainable development (extravert tendency); the second is looking for a measurable and verifiable result in his immediate environment and is not concerned with the global impact of his choice (introvert tendency).

Diversity of the motivation in consumption choices with less ecological impact

One of the most interesting results of our study underlines the diversity of individual motivations that play a role in the consumption choices which exert less impact on the environment and variation of the expressions of these individual motivations for different sectors of consumption.

We have observed that participants do not make 'coherent' choices from an environmental point of view for all the sectors of their consumption, as these choices obey other coherence systems. Thus a person can adopt behaviour and make choices that are environmentally friendly in one sector of consumption without necessarily making the same choices in other sectors. For instance, he or she may buy a washing detergent that obtained an environmental label, strictly respect the recommended dosages and choose water- and energy-saving washing programmes, while, on the other hand, he or she may use highly toxic paints for personal reasons (the colours) and may dispose of used solvents in the toilet.

The need to develop diverse strategies to promote sustainable consumption

Considering the variety of profiles of motivations, it is illusory to think that it is possible to develop a single strategy in the field of sustainable consumption that could be adopted by everyone. On the contrary, developing multiple strategies seems to be necessary to meet the motivations of the various profiles, which may be very different from environmental or altruistic motivations. However, the measures that are currently taken by the public authorities or other actors to encourage changes in consumption behaviour are often based on the presumption of an 'altruistic' motivation in favour of 'environmental protection' which is very general and sufficiently strong, and which could set off behavioural changes when stimulated correctly. In the light of our results, it may thus be argued that to date the talk of sustainable development appeals to the principle of solidarity and sharing of space and time, an approach that mainly inspires people whose profiles show merging introversion and more generally belonging to the group (but a planetary group in this case). Thus we could say that this approach only interests persons with certain profiles, while others are potentially interested in ethical or ecological consumption choices but do not find levers for their motivations in this approach.

Sectoral approach: the case of the laundry

Among all the different sectors we examined, we give the example of the laundry below, showing how individual motivations intervene in the determination of choices, for the detergent as well as for the appliances (washing machine and drier).

Motivation profiles

Not all the participants have the same objectives when they do the laundry. The objectives are dependent on lifestyle and some socio-demographic criteria (e.g. family, single person), but they are also indicators of profound motivations related to personal psychological tendencies. The analysis shows that two main perception levels play a role in laundry: priority is given either to the task or to the result. Participants who give priority to the task mention the 'fatigue' aspect associated with sorting beforehand, drying (e.g. hanging on clothes-line), ironing and storing, and the 'formality' aspect that has to do with the inevitable, repetitive, routine nature of the laundry chore. Those who give priority to the results refer to the satisfaction and pleasure that comes from the result: cleanliness, nice smell, bright colours, the feeling of bringing pleasure and well-being to those for whom the clean laundry is intended. Here we have disclosed six major objectives, each leading to different approaches to laundry.

- *Profile 1: Speed* (priority given to task). These persons are active and determined, looking for innovation. They consider laundry to be a chore and want to devote as little time to it as possible. Every solution to speed up or facilitate the laundry will be applied, as long as it is innovating and individual. For these persons, ideally, laundry should be done entirely by the machine.

- *Profile 2: Bright colours and sparkling white* (priority given to result). These people seek a certain standing; they worry about their image. Doing the laundry is seen as a chore and they do not want to spend time on it. They do the laundry (i.e. buy laundry products and use the washing programme) to obtain 'perfect' results in terms of bright colours and sparkling white.
- *Profile 3: Cleanliness and cost-effectiveness* (priority given to result). These people want absolute cleanliness. They see laundry as a moment to put things back in good order: they manage the laundry to recover control of a situation that was disturbed by spots on the fabrics. They are willing to invest time rather than technique to achieve their objective. Ideal laundry is a totally controllable result.
- *Profile 4: Sweet smell and soft touch* (priority given to result). These people have a tendency to be maternal and protective. They consider laundry to be a pleasant task, put time into it and do not mind doing the laundry every day. The smell of clean laundry gives them pleasure and they enjoy the sensation of well-being it gives their loved ones. These are the people who prove to be the most concerned about the environment when it comes to doing the laundry; they find that whatever endangers human health can also damage the environment, and vice versa.
- *Profile 5: Basic cleanliness or 'no stains'* (priority given to task). These people look for cleanliness that is acceptable to everyone. They do not demand absolutely clean clothes, but they cannot be dirty. They prove respectful of other people and want to live in harmony with others. They see the laundry as a formality or an obligation.
- *Profile 6: Facility* (priority given to task). These people are carefree, not under stress and like to give themselves a little pleasure. They love company and contacts. They see the laundry as a 'bother' and only do it when it is absolutely necessary. They try to simplify tasks, but do so by sharing a task rather than by applying innovating technological solutions. Ideally, they would like to do their laundry along with several people by sharing the tasks to reduce the 'chore' aspect and the risk of 'accidents' (e.g. shrinking, fading of colours).

In fact, the reality is more complex: an individual rarely corresponds perfectly to any single tendency described above. In general people develop mixed approaches: one priority for the working method (e.g. speed) can be combined with another priority for results (e.g. bright colours). Nevertheless, when a dilemma occurs, the main tendency overrides (people choose either the result or the working method when both objectives cannot be met). In the laundry world, the preservation of the environment is not such a big priority for the respondents, and certainly not of sufficient importance to compete with their priority objectives; care for the environment can only be taken into account in a complementary way.

Seeking acceptable change scenarios

We submitted a series of suggestions to the participants that could reduce the environmental impact of the laundry and asked them to select those they find feasible or are already applying (Table 6.1).

Table 6.1 Laundry case: seeking acceptable change scenarios

	1 Speed	2 Sharp colours	3 Cleanliness	4 Pleasant fragrance	5 Basic cleanliness	6 Facility
Choose clothes and textiles that do not soil easily, that are easy to clean		–				+
Choose a washing machine with the label AAA (water- and energy-saving)	+	+	–/+	+	+	+
Change in the way a person does the laundry	–	–	–/+	–	–	–
Buy detergents that are more respectful of the environment	+			+	+	+
Buy a drier with low energy consumption or do without a drier				+		+
Do nothing	–	–				–
Combine several possibilities				+	+	+

The suggestions were received differently in view of the motivation profiles. In general, participants agree more easily to change their purchases and product choices (as to appliances and washing products) than to adapt the way they do the laundry. The economic criteria are important to all the participants and set the limits of the possible purchases of machines and detergents. However, the economic criterion is relative: although all respondents put limits on their expenditure for the laundry, these limits vary with the motivation profiles within a category of consumers who have more or less equal incomes or similar lifestyles. We already underlined that, for the participants, products should only be allowed on to the market when they do not present any danger whatsoever for the environment. They find that the responsibility for the products lies with the producers and that it is up to the public authorities to impose standards and rules, to ensure protection of the environment. The participants feel incapable of evaluating the environmental quality of a product and systematically underestimate the impact of their actions. The participants' expectations with regard to the products go further than mere respect for the environment; they expect products and also the machines to provide a sort of 'education' about more eco-friendly behaviour, by interactively providing them with advice for use, or by informing them in real time of water and energy consumption among other things.

Measures promoting an ecological choice of a washing machine and a detergent

We have tested different propositions of measures that can convince consumers to take the environment into account when they purchase a washing machine or a detergent. For washing machines and detergents, the elements that could influence the choice vary with the motivation profiles. For washing machines, the proposal that attracts the most interest of a majority of profiles is the one about the energy label. On the other hand, the ecological label is not appreciated and is mainly unknown and misunderstood, which is not surprising on realizing that this type of label is rare on the Belgian market. With regard to detergents, it seems possible that people in all the different profiles would choose an environmental detergent but this choice cannot be stimulated in the same manner for all of them. Nevertheless, some options seem to convince more profiles than others (as long as the prices charged remain similar to today's prices). The 'best' scenario prescribes the launch of a diversified offer of ecological detergents from different well-known brands, supported by promotional actions, particularly test samples. Recommendations made by a close acquaintance or a trusted professional would increase potential adherence.

Changing working methods

We asked the participants to give their opinion about a series of suggestions to get them to change their habits and working methods. Spontaneously, respondents declared that they are less inclined to adopt changes in their working methods than in their purchase habits. They have good justifications to explain why they use a high temperature programme or why they 'need' to wash small quantities of clothes. However, when adopting a series of actions is suggested to them to preserve the environment, the majority of respondents declare that they already act that way (Table 6.2).

Credibility of the sources of information

In the field of sustainable consumption, the programmes of action adopted by public authorities give a considerable place to information measures. In this context, it is interesting to identify the actors in which consumers have confidence. We asked the participants what sources of information would be

Table 6.2 Laundry case: changing working methods

	1 Speed	2 Sharp colours	3 Cleanliness	4 Pleasant fragrance	5 Basic cleanliness	6 Facility
Install the appliances correctly				−		
Fill the machine completely	+	−	+	−	+	+
Sort the laundry by type of fabric, colour, how soiled it is	+	+	+	+	+	+
Choose a lower temperature washing programme	−	−	−	−	−	
Cancel the pre-wash	−					−
Respect the recommended doses	+	+	+	+	+	+
Dry in open air rather than in a drier						
Leave clothes that need ironing out of the drier	+	−	−	+	+	+
Give regular maintenance to the appliances	−	−	−	−	−	−
Minimize ironing	+	−	+	−	+	+

sufficient to convince them to really change their way of doing the laundry (Table 6.3).

The participants give different opinions about the information they seek or take into account before making their choices. In general, they say that they pay little attention to product labels. Some participants, particularly those who have adopted ecological behaviours, feel sufficiently informed and describe various ways of looking up information. Others find that the information they are given is insufficient to allow them to make ecological choices, but do not take into account the information provided by labels, for instance. The problems they mention are not only about the lack of information, but also about the difficulty of recognizing trustworthy advice from among all the information they receive, their confidence in the provider, the time spent looking for information, and their lack of interest in this type of information. Information sources seem to acquire credibility only through their personal competence and proximity. The preferred source of information will be a personal close acquaintance or a person who is recognized as competent (general

Table 6.3 Laundry case: credibility of the sources of information

	1 Speed	2 Sharp colours	3 Cleanliness	4 Pleasant fragrance	5 Basic cleanliness	6 Facility
Advice given by the sales person when the purchase is made				+	+	
Advice given in the manuals that accompany the washing machine			−	+		−
Advice posted on the machine (e.g. stickers)	+	−	+	−	+	+
Advice printed on the detergent packaging	−	−	−	−	−	−
Information about energy their appliances consume	+		+	+	+	
Information about energy consumption on the invoice	−		+	+	+	−
Advice given by relatives and friends	+			+	+	+
Advice given by consumer associations			+	+	+	
Information campaigns organized by public authorities or NGOs				+	+	
Television programmes			+	+	+	
Messages spoken by famous persons	−	−	−	−	−	−
Advice given by a trusted person (e.g. general practitioner)		+	+	+		
The presence of a sick person in the vicinity				+	+	

practitioner, dentist, engineer). At a political level, the closest public authority (the commune) is considered more credible than a more distant public authority, such as the federal government or the European authorities. There is no source of information that is considered trustworthy by all the different profiles. Thus information measures or campaigns should go through different channels. The proximity of the source of the advice (family, friends, doctors), the presence of information on the machines or information displays (e.g. LCD screen) built into the machines and, to a lesser degree, the advice from consumer associations, are all sources that can encourage voluntary adherence.

Discussion

The qualitative approach does not give quantitative results. However, by including various socio-demographic factors and motivation analyses, it could highlight a series of options that could meet consumers' various needs.

The probability of adopting more ecological consumption choices in the field of laundry depends on different criteria with regard to the socio-demographic aspect (e.g. composition and size of the family, professional occupation, income, education) but also with regard to the psychological motivations of people.

The options mentioned most frequently by the participants are the introduction of compulsory environmental standards for the production of detergents and the supply of more environmentally friendly products (appliances and detergents), in a variety of brands and distributed in various shops (diversity and extent of supply). Consumers want the public authorities and producers to take the initiative; they see themselves more as recipients than instigators of initiatives. Choosing eco-friendly washing machines seems to be the most promising option for consumer acceptance. The competition between several brands could stimulate acceptance even more. The energy label[10] seems to be the most efficient vector for that purpose, but its presentation and comprehension should be evaluated accurately to optimize its impact. In that field, advice and recommendations coming from close acquaintances are more effective than any other argument, the sales person included. Choosing an eco-friendly detergent could gain major potential acceptance. The most promising scenario in that context is the introduction of several detergents on to the market, made by 'well-known' brands respecting ecological criteria guaranteed by official standards. A European ecolabel[11] that is visible and compulsory could have some success, on condition that it is made more visible. Indeed, participants declare that they do not read the labels on detergents, at least not before actually purchasing them. Test offers and samples seem to be effective instruments to encourage the adoption of new detergents, but the detergent must demonstrate its efficiency and meet other requirements, for instance, with regard to price or fragrance.

It seems that an increase in the supply of ecological products could develop interest of various profiles in this kind of product, not only because a consumer would have a better 'chance' to encounter an ecological product when running errands, but also because the proximity of ecological products seems to be an important source of learning about that question. Nilsson *et al.* (1999) suggest that ecolabels and ecolabelled products, plus their immediate surroundings in shops, are the most important sources of learning and thus of the availability of ecolabelled products in the shops. Broader availability of ecologically differentiated products in shops could well be an important lever to increase the attention that is paid to ecolabels. Moreover, the increasing supply would take ecological products out of the 'marginalization', a fundamental handicap that has a restraining influence on all profiles, except the profile that is looking to distinguish itself from the others through its consumption choices. In that context, the adherence to famous brands would certainly be amplified.

Conclusions

We have used laundry as an example to show that the adoption of ecological behaviour is the result of a complex dynamic that varies from one individual to another and in view of the circumstances (the same individual may adopt different strategies). We could just as well have taken the other examples – paints and clothes – and had the same discussion.

We studied motivations that could explain the adoption of environmental choices. We found that these motivations are diverse, vary from one individual to another and are based on profound psychological motivations. Moreover, these motivations in one person do not manifest themselves in the same way in the different sectors of consumption.

Very few people make consumption choices according priority to environmental protection. At best, they take environmental protection into account in their selection criteria when choosing from several options that would meet their priority motivations.

All the participants say they are aware of the environmental impact of Western consumption patterns, but they feel they do not have the power to change that situation. Various reasons are put forward: the weight of the individual compulsion to explain adopted behaviour; the people see themselves as 'captive' in that they do not have the means to act differently; people do not identify their consumption as an (effective) environmental strategy but find that responsibility for products lies with the producers; the concept of an 'ecological product' is hazy and misunderstood; fear of proximity effects are linked to ethical consumption choices and so on.

The lack of information is not put forward in the same way by all the participants: those who adopt ecological choices and behaviour say they are sufficiently informed and describe various ways of looking up information; those who find themselves poorly informed do not actively seek information and do not use the provided information (for instance, on the product labels).

The participants see respect for the environment as a constraint, an obligation and not as a desire. Consequently, when they face a variety of ecological options, they only adopt the one option that is the least restrictive for them and that does not go counter to their personal priorities. When ecological behaviour is adopted, the underlying motivations often do not fall under environmental concern but under other fields: savings and control, health and security, belonging to the group, a need to feel distinctive and so on.

Changes in product choices and the adoption of more ecological habits are conceivable for a number of participants in the discussion groups, but the envisaged changes and the measures to which they are sensitive vary among other things with their psychological motivations, which also differ from one consumption sector to another.

Given the diversity and variation of individual motivations that intervene in consumption choices, those who have to find ways to encourage changes in consumption habits should be prudent. Indeed, initiatives that seem to present good potential for one category of products will not necessarily be equally advantageous for another category. For instance: the influence of the sales person who gives environmental advice would have a positive effect for several different profiles of consumers in the field of paints, but it would have the opposite effect in the field of washing detergents and a mixed effect in the field of washing machines. On the other hand, the advice of a close acquaintance would be positive for a larger number of profiles in the various product categories that were examined. It is illusory to think that one strategy based on similar measures, or one speech, could influence a majority of profiles and contribute to the development of more sustainable consumption habits. From a strategic point of view, it seems preferable to identify the motivations that support the various attitudes identified as beneficial to sustainable development, and to try to develop these attitudes by using the varied motivations behind them, rather than trying to promote the concept of 'sustainable development' as a motivation to promote certain consumption attitudes. However, if significant changes are needed urgently, then the most effective option will surely be to enact 'compulsory' policies, because changes due to adjustments in individual consumption choices could take time before their effects are perceptible. Indeed, for every field of consumption studied, it never seems possible for the consumer to act as an instigator of sufficient change for the researchers to observe the effects at the society level. On the other hand, there are sectors for which the perspective of potential change is more promising, even if the fields examined offer only limited potential, at least for the near future.

Notes

1 Joint project by the IGEAT (ULB), the Centre for Research and Information of the Consumer Organizations (CRIOC) and the *Centre Entreprise-Environnement* (UCL), in the framework of the Programme for Scientific Support to Sustainable Development (PADD II) under the Belgian federal government's scientific policy. http://www.belspo.be (project CP17).
2 Brussels, June and October 2002.
3 Cf. Chapter 10 (this volume) by Coline Ruwet who studies an already convinced public.
4 PADD II of the Belgian Science Policy.
5 Brussels, 2003.
6 By a 'sector of consumption' we mean a group of products, services and practices that contribute to a specific function: for instance, the laundry sector includes all appliances (washing machine, dryer, iron), products (washing detergent, softener) and practices that allow the maintenance of clothes and linen.
7 Brussels, 2003.

8 C-G. Jung (1950) *Types psychologiques*. Geneva: Georg Editeur, pp. 323–507.
9 This observation is of a qualitative nature and does not pretend to have any statistical validity.
10 Council Directive 92/75/EEC of 22 September 1992 on the indication by labelling and standard product information of the consumption of energy and other resources by household appliances (Official Journal L 297 of 13 October 1992).
11 Regulation on European ecolabel (EC) No. 1980/2000, available online   at http://europa.eu.int/comm/environment/ecolabel/index_en.htm.

7   Greening some consumption behaviours

Do new routines require agency and reflexivity?

Françoise Bartiaux

During the mid-1990s, most Belgian municipalities and communes encouraged citizens to sort their household waste by providing them with new and varied facilities: collection of recyclables, bottle banks, garbage scales, sensitization campaigns and so on. The change in behaviours was massive and rapid: for example, in 2002, the rate of recycling of packaging waste was as high as 70 per cent in Belgium, while the mean for the fifteen former European Union member states was 54 per cent (EEA, 2005).[1]

The scope of this chapter is to analyse this behavioural change and its rationale by comparing household waste-sorting practices with other habits and criteria that are related to green shopping. Results obtained for Belgium are compared with results from other European countries, whether they come from research on household waste sorting and green shopping or from studies on other aspects of green consumerism such as shifting to organically grown food products or saving energy at home.

This contribution begins with a short review of concepts and theories used to understand or predict changes in everyday life practices in an environmentally friendlier way, such as the rational action theory as well as sociological critiques of this theory. Results on changing routines in two areas – household waste-sorting and green buying – are then presented for Belgium,[2] compared and discussed, to sustain this alternative approach with more empirical evidence.

Conceptualizing changes in consumption practices

With which conceptual frameworks is research studying how social or political pressures bring about changes in consumption behaviours in an environmentally sounder way? These studies often carry several assumptions that Macnaghten and Urry (1998) uncover thanks to a critical analysis, which is first summarized. We then discuss these assumptions one by one, focusing on the study of the changes in behaviours.

Common assumptions in sustainable consumption studies

Analysing how 'public bodies are seeking to implicate people in contributing towards a more sustainable future', Macnaghten and Urry (1998: 217–218) outline three dominant assumptions that 'can be read as part of a modernist tradition in which the limits of "natural" processes can be defined unproblematically by science, where public policy and global management strategies can derive from scientific understanding, and where such understandings can engage and mobilise the wider public – the combination of which lead to the ultimate goal of sustainable development'. These modernist assumptions are:

• Nature is conceived as a set of issues identified through modern scientific inquiry . . .; in particular, environmental issues are recognized first and foremost as global/technical issues;
• People are presented as individual agents acting 'rationally' in response to information made available to them. Ignorance about environmental issues can be rectified by the provision of information; information will engender concern; and concern will translate into both personal and political behaviour changes . . .;
• Sustainable development relies on an optimistic model of personal 'agency' . . . people's actions are governed more or less straightforwardly by their knowledge and concern about environmental issues . . . the institutional context in which their behaviour occurs is implicitly assumed to be benign or irrelevant.

(Macnaghten and Urry, 1998: 217–218)

These three assumptions are discussed below with concepts and theories drawn from the literature on changing consumption behaviours in an environmentally friendlier way. In the main part of this chapter, several findings are presented and they demonstrate that these assumptions are not supported by empirical evidence.

Assumption 1: the conception of nature

The world socialization scheme of most Western societies is based on resources predation (Descola, 1999). According to this author, a 'world socialization scheme' is an ethical horizon that orients both the relationships to the environment and the conception of social links and exchange. From his study of Amazon tribes, Descola distinguishes three schemes: predation, reciprocity and giving. In our Western societies, 'nature has become the simple object of our indifferent predations' and this 'world socialization scheme' is made acceptable only thanks to a strong division 'between humans and non-humans' (Descola, 1999: 128).

This radical separation is echoed by the human exemption paradigm, which makes us consider that our human societies, especially in developed countries, are the only ones, thanks to culture and technique, to exempt themselves from the constraints that nature opposes to social activities, a paradigm made clear by several environment sociologists (Vaillancourt, 1996; Macnaghten and Urry, 1998).

In another publication (Bartiaux, 2005), I argue that this division between humans and non-humans is also mental, and that this division may appear to be diffracted into numerous rationales that underlie specific practices, which thus seem to the observer to be uncoordinated with respect to their environmental consequences. Examples may be found in the field of energy consumption, where numerous studies, summarized by Lutzenhiser (1993) and Shove *et al.* (1998), show that consumers are not often aware of the energy consumption that is required in their energy-related practices, either for bathing, communicating (Gram-Hanssen, 2005) or doing the house chores (Bartiaux, 2003). This mental compartmentalization is observable even with individuals or families with a strong environmental concern who have changed several aspects of their daily practices (but not in communicating practices: Gram-Hanssen, 2005) or with those whose profession is related to energy savings. Iversen (1996) and Halkier (2001) also use this concept of mental compartmentalization for characterizing consumption practices that are not related to (other) environmental considerations.

Assumption 2: the rational actor

The second assumption on rational agents echoes quite well the paradigm of rational action, which is best exemplified by two leading theories in social psychology: the theory of reasoned action (Ajzen and Fishbein, 1980) and the theory of planned behaviour (Ajzen, 1985). In summary, both theories predict behaviour by the intention to do this behaviour and by the attitudes towards it; in turn, the attitudes are predicted by personal beliefs about the consequences of the behaviour and the evaluation of those consequences, while the intention results from these attitudes and the subjective norms, which are themselves explained by the beliefs in the expectations from significant others about this behaviour as well as by the motivation to comply and meet these expectations. Finally, and this is the addition of the theory of planned behaviour, both the intention and the behaviour itself are also determined by the perceived control over the obstacles and over facilitating elements.

By placing individual beliefs at the first step of the causal model of behaviour, and especially the beliefs about the consequences of the behaviour, both theories award a central role to information that can modify these beliefs and consequently the attitude and the intention to behave. However, several sociological studies on changing behaviours in an environmentally friendlier way have demonstrated that individuals are not simply taking in new information or environmentally oriented advice as such: on the contrary, they interpret this information. Researchers have pointed at several conditions: the necessity of bringing new knowledge from practical (and 'hidden') to discursive consciousness (Hobson, 2003; following Giddens, 1984), the required 'convergence and not the quantity of knowledge that determines its effectiveness in promoting ecological behaviour' (Goldblatt, 2003) as well as the important role in these respects played by the individuals' social networks – either real, with familiar persons, or virtual, via the mass media – (Gram-Hanssen *et al.*, forthcoming). According to them,

> social support is indeed necessary to make changes possible: 'the demand for acknowledgement overcomes society. Everyone is on the lookout for approval admiration, and love in the eyes of the others. . . . Without limitations. . . . Self-esteem is at the origin of every change' (Kaufmann, 2004), as identity management is associated with social support (Caradec and Martucelli, 2004).

The necessity of identity management appears indeed to be a powerful explanatory factor of behavioural change, even when it comes to household chores and daily routines (Kaufmann, 1992, 1993, 1997).

Assumption 3: agents behaving irrespective of the institutional context

This third assumption on personal agency is consistent with the second one on rational actors. In this hypothesis, the agency is given such a high importance for changing behaviours in an environmentally friendlier way that it seems to overcome any contextual factor, such as institutions, facilities or social norms. As pointed out by Macnaghten and Urry (1998), this appears to be too 'optimistic' a view of individuals, having the power of 'changing the environment fate' (Shove, 2003) by their behaviour.

In the following part of this chapter, several analyses demonstrate that these three assumptions are not supported by empirical evidence, in particular when it comes to the rationality of consumers and the benign role of the institutional context.

New routines: a result of a self-reflexive attitude?

Data and methods

This research is based on a quantitative survey that was performed in 1998 on a representative sample of about 3,780 households in Belgium, on the one hand (this survey is known as the Panel Study of Belgian Households [PSBH]). On the other hand, this contribution is also based on some fifty in-depth interviews (following the method of Kaufmann, 1996) which were conducted from 1997 to 1999, most often with couples, some with persons living alone and others with parents and children.

Two scores of environmentally friendlier actions are calculated: the first is related to the practices reported by the respondents on household waste-sorting (frequency and range of materials sorted); the second score relates to the practices and the criteria that the respondents have for their grocery shopping.[3] Both scores have a maximum of four. They will be computed and compared for various categories of households, as explained below.

Constructing new routines

The following quotes set the stage by pointing out the difficulty in changing habits and thinking about practices that are normally routines, the fear of novelty or of not being able to do it, the necessity to – literally speaking – incorporate new gestures as well as, after some time, the easiness of new routines and some pride in having overcome this change: 'It was the unknown! Yes. It is stupid, isn't it? It messes up habits, also. But for me, it wasn't at all . . . Again new things! But no: it's for the better! It was the fact of beginning [that was difficult]. But now, there is no problem!' (Anita). '[My first reaction was that] I was bored! . . . Inconvenients are really minor according to the advantages that one can get from it. At one point, we thought that it would be beyond our competence but . . . now, it is incorporated into habits! Well, I speak about us. We are not even thinking of it anymore! I think we systematically do it and it doesn't bother me. Honestly, it doesn't bother me. I don't consider it an inconvenience' (Daniel).

The role of political pressure

According to our Belgian survey, the perceived pressure to sort household waste is related to both the frequency and the range of the types of waste sorted (the Pearson's correlation coefficient between the score of perceived pressure and the score of sorting is quite high: $R^2 = 0.40$). When the respondents perceive a weak pressure to sort their household waste, they sort it all the more if they have a higher level of education. In Denmark, Jensen (1996) has also noted variations in the types of sorted waste and in the amount of residual waste according to neighbourhoods, which for him denotes socio-economic differences in standards of living. However, and back to Belgium, when a strong pressure is perceived, all households have the same score on sorting, whatever their education level (Table 7.1). Thus, the perceived external pressure plays an important role in influencing *all* households (whatever their socio-economic status) to sort their waste while at the same time the facilities provided, which are indissociable from this obligation, are certainly also encouraging them to do so.

In Holland, there have been a number of experiments with financial instruments based on the 'polluter pays' principle. In the town of Oostzaan, for example, households were charged according to the amount of collected waste. Within the first year, the amount of waste requiring collection fell by 38 per cent (Linderhof and Kooreman, 1998).

Table 7.1 Waste-sorting score according to the level of education and of perceived external pressure (mean scores out of 4)

Education level	Weak pressure		Strong pressure		Total		Perceived external pressure	
	Sorting score	N	Sorting score	N	Sorting score	N	Pressure score	N
Primary	2.25	258	3.61	728	3.25	968	3.11	968
Low secondary	2.39	402	3.64	1,216	3.34	1,605	3.14	1,605
High secondary	2.60	468	3.68	1,497	3.43	1,940	3.17	1,940
Higher education	2.70	539	3.68	1,344	3.40	1,855	3.05	1,855

Source: all the respondents, PSBH (1998) (N = 6840).

In the field of energy consumption, the importance of the context and especially the differences of energy policies carried on – or not – by Denmark and Belgium were also found to be quite useful in explaining the higher electricity consumption of Belgian households than of Danish ones; furthermore, these differences in energy policies may also be related to some differences in behaviours or proportions of households owning specific appliances in the two countries (Bartiaux and Gram-Hanssen, 2005).

The institutional context in which individuals operate is thus quite important and must be taken into account in the analysis, as recommended by Macnaghten and Urry (1998) and contrary to the third assumption, mentioned above, that they have often found in the literature.

The role of self-evaluation as a polluter or not

Another dimension of social representations on environment is the self-evaluation as a polluter or not, which was expressed by the following question: 'Do you consider yourself as someone who pollutes the environment?' This question comes from a French survey (Collomb and Guérin-Pace, 1998: 241). As in France, the majority of answers in Belgium are: 'Yes, as most people.' However, the proportions for this answer are quite different in both countries and it is difficult to propose a satisfactory interpretation of this difference (Table 7.2).

The best sorters are to be found among those who 'consider themselves as polluting the environment less than most people', which shows a consistency between self-evaluation and practice (Table 7.3). However, their score of sorting is nearly as high as the one obtained by the respondents who do not 'consider themselves as polluting the environment': do they grant themselves this positive evaluation thanks to their good practice in sorting waste, even if their score on 'green buying' is the lowest? Perhaps they think that this awareness when shopping only concerns other people: those who pollute the environment. In

Table 7.2 Self-evaluation as someone who pollutes the environment (France and in Belgium)

Do you consider yourself as someone who pollutes the environment?	Belgium (%)	France (%)
No	30.0	20.0
Yes, but less than most people	31.0	23.0
Yes, as most people	38.6	55.8
Yes, more than most people	0.4	0.6
Total	100.0	99.4[a]

Source: all the respondents, PSBH (1998) (N = 6840).
Source for France: Collomb and Guérin-Pace (1998: 241).

Note: [a] The French survey also included the categories 'Does not want to answer' (0.2 per cent) and 'no answer or does not know' (0.4 per cent).

Table 7.3 Mean score of environmentally friendlier action according to self-evaluation

Do you consider yourself as someone who pollutes the environment?	Household waste sorting (score out of 4)	'Green buying' practices and criteria (score out of 4)
No	3.42	1.85
Yes, but less than most people	3.44	2.06
Yes, as most people	3.27	1.87
Yes, more than most people	2.89	1.89
Total	3.37	1.92

Source: all the respondents, PSBH (1998).

another publication (Bartiaux, 2002: 146–147), we have shown by several examples that in couples and families, some members are addressing silent injunctions to others in order to get him or her to do some task that they consider degrading.

Thus rationality and actors' strategy may sometimes consist in saying: 'Do yourself what I am preaching (or wishing for) and not doing!' The rational–actor paradigm does not take this kind of rationality into account, since it only considers the individuals' networks as motivators, not as surrogate actors.

Self-evaluation for being a polluter does not completely predict environmentally friendlier actions, especially when it comes to green shopping. These results also show that in Belgium in 1998, grocery shopping was unrelated to pollution and environmental considerations for 70 per cent of the people. In another publication based on more recent (2004) qualitative data (Bartiaux,

2005), I argue that acknowledging the systemic character of environmental matters seems to be a threshold and, when it is not acknowledged, consumers justify their behaviours with what may appear to an external observer as fragmented rationales.

The plurality of action rationales

Following Boltanski and Thévenot (1991), we have tried to use in our survey the different 'cities' and their justifications to get an indication on how people comprehend and problematize waste issues and potential solutions to them by adding the following question: 'According to you, what are the two principal solutions to solve the problems raised by waste?' and by proposing several 'solutions', which are based on the justifications shown by these authors:

- The civic justification based on collective involvement and equality ('By actions taken by the inhabitants in each neighbourhood').
- The industrial justification based on efficiency ('By improving industrial production systems').
- The domestic justification based on interpersonal relationships ('By actions taken in each household').
- The justification through opinions based on other people's acknowledgement ('By sensitization campaigns').
- The economic justification based on the market ('By developing the economic value of waste').
- The inspired justification based on creativity ('With imagination').
  We have added three further justifications:
- The ecological justification based on a general change (Lafaye and Thévenot, 1993) ('By changing lifestyles').
- The political justification based on delegation ('By actions taken by public authorities').
- The international justification based on world conferences and events ('By actions taken by international organizations').

According to the answers to our survey, the best way to solve waste problems would be the ecological 'solution' (30 per cent) because, or in spite, of its vague character ('By changing lifestyles'). Other less often cited justifications include the domestic one, the industrial one and the economic one (about 15 per cent each). These last two justifications are most often cited as second choice (19 and 18 per cent, Table 7.4).

If we assume that the domestic answer denotes a sense of agency (a feeling of being able to act usefully – for example, for the environment) the result that 15 per cent of the Belgian population choose this answer tends to confirm the findings of Macnaghten and Urry for the United Kingdom: these authors have shown (by qualitative methods) mixed attitudes on personal agency with a strong mistrust towards institutions that should manage environmental changes

Table 7.4 Type of justification invoked to solve waste problems

In your opinion, what are the two main solutions to solve waste problems? They can be solved . . . (N = 6742)	*First choice*	*Second choice*	*Non-weighted total*
By changing lifestyles	30.0	13.2	43.3
By developing the economic value of waste	13.8	18.0	31.8
By improving industrial production systems	14.5	16.9	31.4
By actions taken in each household	15.1	13.4	28.5
By actions taken by public authorities	11.2	13.2	24.4
By sensitization campaigns	3.8	9.1	12.9
By actions taken by the inhabitants in each neighbourhood	2.5	5.7	8.2
With imagination	4.1	2.9	7.0
By actions taken by international organizations	5.0	7.6	12.6

Source: all the respondents, PSBH (1998).

(1998: 246). They thus demonstrate that their results do not confirm the implicit assumption that 'sustainable development is based on a model of individual agencies', a model that they consider as too 'optimistic' (1998: 218).

As shown in Table 7.5, the solutions to solve waste problems according to the survey respondents are associated with the environmental actions they have done or have not done: those who answer 'By actions taken in each household' are the best sorters as well as those who choose the political solution, 'By actions taken by public authorities', which again shows the influence of social pressure on household waste-sorting. Besides, the respondents who choose the ecological 'solution', 'By changing lifestyles', are those who pay the most attention to environmental criteria when they do their grocery shopping; they are followed in this matter by the respondents who propose the domestic 'solution', 'By actions taken in each household'. The results for the domestic 'solution' give some more credit to the hypothesis according to which this 'solution' is a good proxy for the agency feeling.

The plurality of action rationales should also draw our attention to the need to avoid interpreting actions as 'environmental' actions that the consumers purposefully realized, for these actions may be defined quite differently by their authors. This shows the interest of linking a qualitative research to a quantitative survey. For example, a woman buys refills for packs of detergent because 'it's more convenient'. Another buys bottled water from her neighbour's shop, not for ecological reasons (less waste and less transport) but for 'good neighbourhood relationships'. These examples and many others show the necessity to split the action–rationality paradigm because the rationales are numerous. According to Bourdieu, there would be more 'intellectualist' situations and

Table 7.5 Mean score of environmentally friendlier actions according to the type of
justification invoked to solve waste problems

In your opinion, what are the two main solutions to solve waste problems? They can be solved . . . (N = 6742)	*Household waste sorting (score out of 4)*	*'Green buying' practices and criteria (score out of 4)*
By actions taken in each household	3.55	1.93
By actions taken by public authorities	3.41	1.84
By changing lifestyles	3.37	1.98
By developing the economic value of waste	3.37	1.93
By improving industrial production systems	3.34	1.92
By actions taken by the inhabitants in each neighbourhood	3.31	1.90
By actions taken by international organizations	3.26	1.91
By sensitization campaigns	3.13	1.78
With imagination	2.98	1.82
Total	3.37	1.92
Signification level (F)	0.0001	0.0001

Source: all the respondents, PSBH (1998).

more 'practical' situations. Between these two poles, the presence of reflexivity during action would be variable (Corcuff, 2002: 70).

The diversity of action involvement patterns should not be mistaken for the identity diversity, even though 'they join up . . . to open new paths towards the study of individuality'; a challenge for sociology, according to Corcuff (2001: 111). In the following section, we will further investigate the notion of identity with a new concept: the secondary and non-conscious benefits. This again shows the inadequacy of the rational–actor paradigm as well as the notion of intentionality of action.

The secondary and non-conscious benefits

The in-depth analysis of the interviews has shown many secondary benefits that are non-conscious for the interviewed persons: a survey by questionnaire could thus not apprehend them, as these benefits are neither rational nor rationalized by the surveyed persons. However, these benefits are real ones, and they contribute to influence the practices of waste sorting. Indeed, to sort and to tidy up as well as to separate oneself from waste are means of identity management and spatial limits definition (Kaufmann, 1997).

These secondary benefits are associated with various identity dimensions: social, parental or conjugal. For instance, for a Spanish immigrant in Brussels, sorting her household waste gives her the benefit of feeling that she integrates herself as a good citizen ('I say that everyone has to make some effort'). A university researcher wants to keep up his social position, and his distinction (Bourdieu, 1979), by applying the new behaviour perfectly ('Since we have to do it anyway, I might as well do it correctly'). Mothers are often trying to defend their territory, which seems to be part of a mother's identity, by taking care of every task themselves, including sorting household waste (this even happens when their children already have a partial residential autonomy: 'they already have to be in charge of their household chores all week, so as they are not totally used to that, it's good for them to have someone else sorting out the leftovers during the weekends'). Secondary benefits can also be conjugal ones, as seen with several couples where the husband has taken over this new chore to counterbalance his weak involvement in other household chores.

For only one female interviewee, there are no secondary benefits and, contrary to all other respondents, she is ready to go back immediately and with no regrets to the previous situation (to the question 'If one was telling you: "It's finished, we stop it", what would you say?', she laughs and quickly answers 'Yes! I stop it!'). This woman is a cleaning lady who earns a low salary. It may be hypothesized that to draw secondary benefits depends on individual reflexive ability, and thus on education level and on social network, but this hypothesis should be further developed and tested.[4] If it were empirically verified, this would mean that an a posteriori rationalization and reflexivity help support newly established routines.

A policy implication of these secondary and unconscious benefits is that the public interest to continue to participate in programmes of household waste sorting is often quite far from the environmental reasons that motivated these policies in the first place, as this interest is here clearly related to identity concerns (see also Rousseau and Bontinckx, Chapter 6, this volume).

No anticipation of sorting practices through grocery shopping criteria

For the defenders of the rational–action paradigm, the most surprising result is perhaps the striking lack of optimization of sorting practices through adequate shopping: the consistency between green grocery shopping and waste sorting is very weak (the correlation coefficient between these two scores is $R^2 = 0.058$). While 66 per cent of the respondents may be considered as fair to very good sorters, the remaining one-third hardly or never sorts waste. But to the question 'Do you buy grocery taking into account the waste generated by what you buy?', only 15.1 per cent answer 'Often' or 'Always', and 57.1 per cent answer 'Never'! Even among the best sorters, the corresponding proportions are not so different: 18.7 per cent and 50.5 per cent. The in-depth interviews show the same mental separation between these two types of practices as shown by the following quotes: 'Quite frankly not. No, we had never thought about it' (Anita). 'It's not a criterion' (Frédéric). 'I'm not going to torture my brains to know what my shopping will produce as waste' (Christine).

Before trying to interpret this mental compartmentalization, let us underline that this process has been noted in other sociological research on environmentally related topics: as shown by Iversen (1996, cited in Halkier, 2001: 37), 'consumption practices are characterized by compartmentalization in relation to environmental consideration'. In research on young Danes consuming organic food, for example, Halkier met a young worker whose dream is to buy a car, and a young lady who is very concerned about her food but 'finds it tiresome to locate a bottle bank'. For Halkier, compartmentalization keeps 'green reflections out of certain practices' and by doing so, makes it possible 'to signal social normality' (2001: 39). Another interpretation, consistent with Halkier's, is that this compartmentalization allows one to avoid adding a new identity dimension whose management would require a supplementary and 'never-ended physical and mental zapping' (Dortier, 1998: 52). Yet this zapping supposedly requires all the more energy that the dimensions to be conciliated are numerous (Bartiaux, 2002), whereas individuals are constantly choosing 'strategies that are cognitively economical' (Kaufmann, 2001: 211; Pacteau, 1999: 336). This mental compartmentalization is another process that invalidates the rational–action paradigm.

Yet another interpretation, I argue elsewhere (Bartiaux, 2005), is that this mental compartmentalization is a societal self-defence mechanism for avoiding a societal acknowledgement of the systemic characteristic of environmental issues. Not acknowledging this systemic character of environmental issues makes it possible to avoid abandoning our world socialization scheme as well as corresponding societal values seen as important in our societies, among which the myth of the mass-consumption society is prominent. This mental compartmentalization may thus be interpreted as a means of avoiding a societal self-deception that would occur if our societies had to abandon its myths and its world socialization scheme.

The role of environmental knowledge

In 1998, in Belgium, only one person out of two knows that the Earth's climate is warming, and one out of three cannot answer the question: 'Do you think that in twenty years the Earth's climate will be: the same, colder, warmer?' More than half of the respondents (52 per cent) answer 'warmer', one-third do not know (32 per cent), 9 per cent think that the Earth's climate will be the same and the others estimate that it will be colder (3 per cent) or give no answer (3 per cent).

As a sociological study conducted in the United Kingdom revealed (Boardman *et al.*, 2003), a majority of people are concerned neither with climate change or environmental issues, and for these authors general awareness of the level of CO_2 emissions is insufficient in Europe. Climate change is an increasing topic in daily citizens' conversation but it is not perceived to be the most important of all environmental problems (Kasemir *et al.*, 2000). People's knowledge on climate change is often confused with other problems such as ozone depletion or pollution, as we have also shown in the case of Denmark (Gram-Hanssen *et al.*, 2005).

Table 7.6 Mean score of environmentally friendlier actions according to climate change knowledge

Climate change knowledge	Household waste sorting (score out of 4)	'Green buying' practices and criteria (score out of 4)
Bad (n = 641)	3.36	1.76
Weak (n = 1869)	3.37	1.88
Rather good (n = 2550)	3.37	1.97
Good (n = 756)	3.36	2.02
Total (n = 5816)	3.37	1.92
Statistical signification (F)	0.9405	0.0001

Source: all the respondents, PSBH (1998).

Household waste-sorting is largely independent of this knowledge whereas green buying practices and criteria are related to it: the better the knowledge, the higher the score (Table 7.6).

In general, and as shown in Table 7.7, respondents who know about global warming have a better knowledge about the factors involved: for example, they are less likely to answer 'I don't know' for each proposed item than in the total sample. However, the differences are weak and there is hardly any difference when it comes to factors that have no effects on climate change. Among the respondents who know about climate change, a certain proportion (45 per cent) think that residential heating is a factor of climate change and the comparison between their answers on the various proposed items enables one to raise the following hypothesis: when the factor is evoking the 'others' or the 'elsewhere' (factories, the Amazon rainforest and maybe car traffic), more respondents point out this factor of global warming than when they are directly concerned with the factor as residential heating. Although this hypothesis should be further tested, it may be interpreted as another example of mental compartmentalization.

Answers given to the questions mentioned in Table 7.7 are combined in a score on climate change: the score varies from 1 to 10. Only 2.1 per cent of the sample answered the eight questions on climate change correctly and they score 10/10. Eleven per cent score 8/10 or 9/10. One-third of the sample has a score of 6/10 or 7/10, and 53 per cent have a score equal to or lower than 5/10. The average score on climate change is 4.9/10.

In Belgium, the score of knowledge on climate change is highest for respondents who answer 'Yes, more than most people'. Does this result suggest the difficulty of translating environmental information into practice, practice that would be reflected in the self-evaluation? Or does it suggest that for these better informed persons, the reflexivity level – in the sense of Beck (1992) – is higher and consequently the self-evaluation is more severe? Results displayed in Table 7.3 tend to support the first interpretation, as both scores on sorting and on shopping were low for these respondents.

Table 7.7 Knowledge on climate change factors

Population	All respondents			Only respondents who think that climate will warm up		
In your opinion, what could modify the climate?	Yes	No	Doesn't know	Yes	No	Doesn't know
Car traffic	65.7	16.8	17.5	77.7	13.6	8.7
Pollution of underground water supplies	43.6	32.5	23.9	48.7	34.9	16.4
Residential heating	36.5	38.7	24.8	45.1	38.5	16.4
Nuclear plants	51.6	24.7	23.7	57.8	26.5	15.7
Dumping dangerous products in landfills	54.5	23.5	22.0	60.0	24.9	15.1
Factories' smoke	73.5	11.4	15.1	82.4	9.6	8.0
Deforestation of the Amazon rainforest	72.0	8.3	19.7	81.8	7.8	10.4

Source: PSBH (1998) (N = 7028).

Note for interpretation: values in grey show the correct answer.

Table 7.8 Mean score of climate change knowledge according to self-evaluation

Do you consider yourself as someone who pollutes the environment?	% (Belgium)	Climate score (out of 10)	% (France)
No	30.0	4.31	20.0
Yes, but less than most people	31.0	5.38	23.0
Yes, as most people	38.6	5.00	55.8
Yes, more than most people	0.4	5.55	0.6
Total	100.0	4.91	99.4[a]

Source: all the respondents, PSBH (1998) (N = 6840).

Note: [a] See note to Table 7.2.

The role of environmental representations

Are green consumption behaviours matching specific representations of the environment? Before answering this question, we first analyse with which word the word 'environment' is associated. A list of eighteen words, presented in alphabetical order, was proposed to the respondents; the word 'nature' was not included after the pre-tests had shown that this word was quoted too often.

The same choice has been made in a French survey (Collomb and Guérin-Pace, 1998: 22).

Table 7.9 shows that for the majority of respondents, the first word associated with 'environment' is 'air', the second is 'water', and the third is 'health'. These three words are quoted more often than the word 'ecosystem'. The analysis of these associations reveals that the choices of the participants to the survey are quite varied: it would thus be an error to think that the word 'environment' is generally associated with its scientific meaning represented here by 'ecosystem'. Another interpretation is that the representation of the environment may be partly considered as an anthropocentric one, as the word 'health' is so often quoted.

It seems consistent that the best score of knowledge on climate change is obtained by the respondents who have selected the word 'ecosystem' to be the most associated with 'environment': their climate score is 5.92/10. They are followed by the respondents who have chosen 'planet' (5.40), then 'water' (5.22), 'vegetation' (5.15), 'future' (5.09), and 'politics' (5.0). The lowest score (4.25) is obtained by the respondents who associate environment with 'countryside' or with 'family' (4.33).

Table 7.9 Words associated with the word 'environment' and climate score

From the following list, choose three words which are in your opinion associated with:	First word associated with environment	Second word associated with environment	Third word associated with environment	Non-weighted total	Climate score (out of 10)
Air	31.4	16.8	10	58.2	4.81
Animals	3.2	6.0	6	15.2	4.60
Future	5.1	5.7	6.6	17.4	5.09
Beauty	1.4	1.8	2.6	5.8	4.64
Calm	2.3	3.1	2.9	8.3	4.59
Countryside	4.3	5.7	5.4	15.4	4.25
Water	8.7	22.7	12	43.4	5.22
Ecosystem	9.0	5.9	7.1	22	5.92
Family	2.8	3.2	2.5	8.5	4.33
Mankind	4.3	5.4	6.1	15.8	4.90
Heritage	0.7	1.3	2.2	4.2	4.56
Planet	6.1	5.1	7.3	18.5	5.40
Politics	0.7	0.9	1.6	3.2	5.00
Present time	0.2	0.3	0.6	1.1	4.92
Health	16.9	9.8	12.8	39.5	4.65
Vegetation	2.0	4.8	9.7	16.5	5.15
Town	0.4	0.4	1.2	2	4.96
Neighbourhood	0.7	1.1	3.4	5.2	4.62

Source: all the respondents, PSBH (1998) (N = 7022).

To summarize all these results, an analysis of multiple correspondences is performed (Figure 7.1). The vertical axis may be interpreted as the external pressure, which is high in the lower part of the graph and low at the top; this axis also shows the voluntary actions as opposed to those done under public constraint. The horizontal axis at the left shows the actions that are friendlier for the environment and at the right the actions that have a negative impact on it. It is possible to define four groups. In the upper right quadrant one notices the group of indifferent people: these people do not feel obliged to sort their household waste and they do not do it; nor do they pay attention to environmental criteria for their grocery shopping. They associate the word environment with words such as countryside, calm and neighbourhood. The lower right quadrant represents refractory persons who are obliged to sort their waste but do not do it; no word appears to characterize this group. The next quadrant, the lower left, brings together the reacting persons: they sort household waste because they feel obliged to do so and sometimes they have 'green' criteria for their grocery shopping; they associate the word environment with 'future', 'air' and 'health'. Finally, the upper left quadrant shows the proactive households (they sort household waste without feeling obliged to do so and they have green

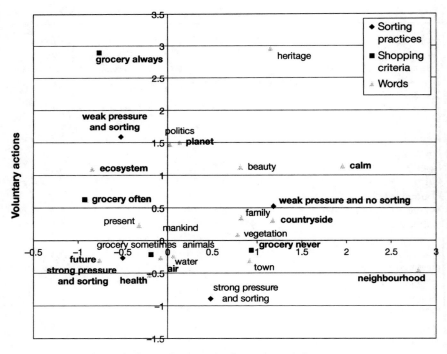

Actions with negative impact on environment

Figure 7.1 Analysis of multiple correspondences on sorting practices groups, shopping criteria groups and words associated with 'environment'

consumerism criteria and practices) for whom 'environment' is related to 'planet' and 'ecosystem'.

The cognitive frameworks thus appear to be continuously adapted according to practical choices, and it seems exaggerated to say that they determine these practices as the influences could be reciprocal. In-depth interviews also show that the demand of identity continuity requires mental arrangements, such as occultation, accentuation and so on. For example, several people have related their new sorting behaviours with their old habits of avoiding throwing papers away without asking themselves why they never made this association before the new policy on household waste. Similar results have been obtained in France by Lhuilier and Cochin (1999).

This association between practices and perceptions is also visible in the perception of political pressure for sorting household waste: on average, this pressure is felt more strongly (3.13 out of 4) by those who have not done an 'easy' environmentally related action (such as read an article on environment or look at a TV programme on this topic) than for those who have done it (3.08 out of 4). The difference is even higher (3.11 against 2.93) when it comes to activities that seem more 'difficult' such as voting for a green party, or participating in a collective action for the environment.

Conclusions

This contribution has shown that shifting to new routines for household waste sorting was successful in Belgium in the mid-1990s, thanks to the perceived pressure as well as the facilities provided. This shows the importance of the institutional and practical context in helping set up new behaviours. As this contextual change brought about conversations with neighbours and a large sensitization, it should be asked whether this feeling of obligation may be assimilated to a modality of discursive consciousness (Giddens, 1984): the shift from practical knowledge to discursive consciousness is indeed the condition to bring about environmentally friendlier actions, according to Hobson (2003). Such a contextual change also provides consumers with social support that makes changes of routines possible, even if the sense of agency is rather low. On the contrary, greening grocery shopping seems to require a strong agency feeling, reflexive capacities and an environmental motivation in the absence of a widely supportive context as, for the majority of the population, grocery shopping appears to be largely disconnected from pollution and environmental concerns.

Environmentally friendlier actions are thus not always led by environmental concerns, the moreso in that they are often consolidated by secondary and non-conscious benefits which are not related to environmental considerations.

The rationale–action paradigm is not supported by empirical evidence, which shows to the contrary silent injunctions to others to act, secondary and non-conscious benefits, no anticipation of sorting practices through adequate shopping and mental compartmentalization towards other environmental issues.

Notes

1   For other European countries, the corresponding recycling rates were the following in 1992 (EEA, 2005): Germany: 74 per cent; Sweden: 65 per cent; the Netherlands and Denmark: 57 per cent; Italy: 51 per cent; Finland: 49 per cent; France: 45 per cent; Spain and the United Kingdom: 44 per cent; Greece: 33 per cent.

2   Some of the results presented here and many others not shown are discussed in further detail in Kestemont *et al.* (2001). This research was supported by the grant HL/DD/021 from the Belgian Science Policy Office and made to M-P. Kestemont, F. Bartiaux, N. Fraselle and V. Yzerbyt, Université Catholique de Louvain, in 1996 to 2000.

3   The score of grocery shopping is based on the answers (never, sometimes, often, always) to the following questions: 'When you shop, do you think about the waste that your buying will generate? Is environment protection a criterion for you when you make your grocery shopping? Do you buy products whose packaging is refundable, e.g. glass bottles? Do you preferably buy items whose producer claims to respect the environment? Do you buy refills when it is possible?'

4   Another hypothesis would be that this cleaning lady finds it tiresome to sort her household waste, since she deals the whole day with garbage and dirtiness.

8 Marketing ethical products

What can we learn from Fair Trade consumer behaviour in Belgium?

Patrick De Pelsmacker, Wim Janssens,
Caroline Mielants and Ellen Sterckx

Introduction: marketing fair-trade products

Marketing is an exchange process in which marketers sell products profitably to a market of (potential) customers who try to satisfy certain needs by buying these products. Marketing management entails the identification of market segments and target markets, the definition of a 'unique place in the mind of customers' (positioning), and a consistent and synergetic deployment of marketing instruments, often referred to as the 'four P's of the marketing mix': product and brand, price, place (distribution) and promotion (communication). The task of a marketing manager is to identify market segments, to define a differentiating and sustainable unique selling proposition (marketing strategy) and to translate this marketing strategy into actionable tactics, i.e. designing products, branding, set an acceptable price, distribute products at the right places in the right way, and promote them adequately.

Both (non-governmental) organizations and companies can sell and/or promote ethical products. Ethical products are products that incorporate a certain ethical value, such as environmental friendliness, biological production, production and distribution in a socially responsible manner (for instance, free of child labour, respecting social rights of workers), or on the basis of Fair Trade. Fair Trade may be defined as an alternative approach of trading partnership that aims for sustainable development of excluded and/or disadvantaged producers in the Third World. It is best known by its most prominent characteristic: paying a fairer (i.e. higher) price to producers than is the case in the free market context. When organizations promote ethical products, they do so by campaigning and lobbying, by means of marketing their own ethical brands, and/or by convincing commercial companies to carry a certification label that is issued and controlled by them. Organizations promoting ethical products can try to impose their priorities outside the free market logic. By means of lobbying they can make their ethical products institutionally more attractive (for instance, by convincing the government to impose an ecological tax and bonus system), or even compulsory (for instance, by means of import restrictions for products without a social label). But they can also convince the customer to buy their products or certified brands using the free market mechanism. In this overview no position is taken on the desirability or even the effectiveness of these basic strategies. It does however start from the point of view of the free market logic: organizations and companies marketing ethical products should know and understand consumer behaviour and how it can be influenced using appropriate marketing strategies and tactics. This consumer behaviour and marketing approach is applied here to Fair Trade products in Belgium. The results of a number of empirical studies on Fair Trade consumer behaviour are reported, and the impact for marketing strategy is highlighted. However, these principles may also be applied to marketing of ethical products in general.

In the next section Fair Trade is further defined and the importance of Fair Trade for world markets in general and for Belgium in particular is highlighted. Next, marketing issues are tackled: the most interesting target groups for Fair Trade products and how these products may be positioned, labelling and branding, pricing, distribution, and information and promotion. In the following section the results of a Belgian study are summarized in which the relative importance of the various instruments of the marketing mix is assessed. The final section offers conclusions and marketing implications.

Fair Trade and Fair Trade sales

Fair Trade is often defined on the basis of its best-known component: fair prices for the products of producers in developing countries. In this context a 'fair price' means a price that is higher than would be the case in a free market situation, and one that enables local producers to develop a sustainable business and to produce in environmentally and socially better conditions. In essence, Fair Trade means buying products from producers in developing countries on terms that are more favourable than free market terms, and to market them in developed countries at an 'ethical price premium' (Bird and Hughes, 1997). This higher price to the consumer is warranted by the higher price that producers receive for their products and by the Fair Trade control mechanisms in the trade channel (see also Le Velly, Chapter 14, this volume).

Fair Trade (2005) estimates that, for instance, in the UK alone 250 products from 370 certified Fair Trade producer organizations are sold, involving 800,000 families and workers and approximately five million people in Africa, Asia and Latin America. EFTA, the European Fair Trade Association, estimates that worldwide more than 800 producer groups are involved and that sales of Fair Trade products exceed 0.5 billion euros and are quickly growing: 22 per cent in 2001 and 2002, and over 42 per cent in 2003. Switzerland and the UK are the largest markets in volume; Belgium, France, Italy and the USA are the fastest growing markets (Worldshops, 2005). According to EFTA (1998), 60 per cent of Fair Trade and other sustainable products are food products, about half of which is coffee. Indeed, coffee is by far the most well-developed ethical market internationally, with an estimated average market share of 1.7 per cent of European coffee sales.[1]

Marketing strategy and tactics

Targeting and positioning

What is the most appropriate target group for Fair Trade products? Market segments may be defined on the basis of different sets of criteria, such as socio-demographic factors, values held by consumers and the stages in the buying process: concern, positive attitudes and buying intentions.

In general, ethical or socially responsible consumers are not easily defined in terms of socio-demographic characteristics, in that various studies show contradictory results. Littrell and Dickson (1999) found that US buyers of ethnic products were highly educated, well-off Caucasian women in their forties. Dickson (2001) found that the degree of importance which consumers attached to no-sweat labels on apparel was not affected by age and income. Most studies conclude that ethical buying behaviour is not influenced by gender (Tsalikis and Ortiz-Buonafina, 1990; Sikula and Costa, 1994; MORI, 2000). Vitell *et al.* (2001) in their study of ethical consumer behaviour found no relation with age and income, but higher education seemed to be correlated with more ethical behaviour. In his extensive literature review, Pirotte (Chapter 9, this volume) concludes that among the Oxfam World Shop customers in Belgium the under-25 and over-45 age groups are over-represented. Roberts (1995) found that people not buying from businesses that discriminated against minority groups or women were mainly women with a median age of 47 and slightly lower incomes, but concluded that demographics were not very significant in identifying the socially responsible consumer. Other studies tended to conclude that the ethical consumer was a person with a relatively high income, education and social status (Roberts, 1996; Carrigan and Attalla, 2001; Maignan and Ferrell, 2001).

More specifically with respect to Fair Trade consumption, in a Belgian study with 808 respondents (De Pelsmacker *et al.*, 2005a) four groups of consumers were identified, based on the importance they attached to various product attributes, such as a Fair Trade label, brand and flavour. Eleven per cent of the respondents belonged to the 'Fair Trade lovers' group, for which the Fair Trade label was by far the most important attribute. Another 40 per cent formed the 'Fair Trade likers' group, for which the Fair Trade attribute was still the most important one, but less prominent than in the 'lovers' group. The other two groups were brand and flavour lovers. In terms of socio-demographic characteristics, the 'likers' group did not differ significantly from the rest of the sample. The 'lovers' group was significantly more male, higher educated and between 31 and 44 years old.

Ethically oriented socio-demographic consumer segments may also be identified on the basis of the various stages in the buying process, i.e. a strong concern for ethical issues, and positive attitudes and intentions. These 'intermediate' factors may be regarded as eventually leading to more ethical buying. In a representative sample of 615 Belgians in which attitudes towards

Fair Trade were studied (De Pelsmacker *et al.*, 2005b), women tended to have a slightly more positive attitude towards Fair Trade than men. They were more aware of the Fair Trade concept, and they were less indifferent towards it. Less educated individuals were more sceptical and indifferent towards Fair Trade than higher educated ones. The older respondents were, the more positive they tended to be towards Fair Trade issues. Inclination to action and concern increased with age. Although lower income groups were more sceptical and showed a higher degree of resignation, they also liked Fair Trade products more than the higher income group.

Based on the two Belgian Fair Trade surveys, one might draw a similar conclusion as in previous international studies on ethical consumption behaviour in general: socio-demographic factors do not seem to be a particularly powerful factor of identification of (potential) ethical or Fair Trade customers. More highly educated individuals appear to be generally more positive than lower educated ones. In terms of the other socio-demographic characteristics, the results are inconclusive. In the De Pelsmacker *et al.* (2005a) study, the 'Fair Trade likers' group do not even differ from the average population profile. Nevertheless, this 'likers' group may be an interesting target group to focus upon. The Fair Trade lovers do not have to be convinced any more, but the basically positive attitude of Fair Trade likers may be turned into actual buying behaviour when the right arguments are used.

Market segments may also be defined on the basis of the values that people feel are important. In the Fair Trade study of De Pelsmacker *et al.* (2005b), the four consumer segments are also defined in terms of the importance they attach to the following values in life: conventionalism, competence, sincerity, idealism and personal gratification. It is remarkable that the four groups differ more in terms of values than with respect to socio-demographic characteristics. Fair Trade lovers and likers are less conventional and more idealistic than flavour and brand lovers, and less interested in personal gratification. One is therefore tempted to conclude that target groups for Fair Trade products should be defined in terms of value propositions rather than demographic variables (see also Rousseau and Bontinckx, Chapter 6, this volume).

After having selected the appropriate target groups a clear USP (unique selling proposition) has to be developed, i.e. 'a unique place in the mind of consumers'. In Table 8.1, based on the De Pelsmacker *et al.* (2005b) survey, the percentage of Belgians that mentions a number of reasons to buy Fair Trade products is given. Based on these results, Fair Trade marketers seem to have a straightforward positioning task. Apparently, people understand what Fair Trade is all about, and react very favourably towards the basic Fair Trade proposition. Fair Trade marketers and organizations should continue to emphasize the traditional and basic Fair Trade propositions, and should keep on appealing to the idealistic nature of potential Fair Trade consumers.

Table 8.1 Main reasons to buy Fair Trade products

Reason	% of respondents
A fair price to farmers in developing countries	63
Safe and honest production processes	56
Retention of dignity and autonomy	52

Branding and labelling

In some cases ethical products are sold carrying an ethical brand (such as Oxfam in the case of Fair Trade); in other cases manufacturer brands or private labels (distribution brands) are sold carrying an ethical label (such as Max Havelaar in the case of Fair Trade).

In Europe alone, 240 different ethical labels and brands are used and, when codes of conduct are included, the number amounts to approximately 800 (Fair Trade, 2004). All these competing labelling programmes, certifications and organizations and their complexity may confuse consumers and undermine credibility (Teisl *et al.*, 1999; Nilsson *et al.*, 2004). For instance, Byrd-Bredbenner and Coltee (2000) in their study of female UK consumers' understanding of EU and US nutrition labels conclude that although consumers are able to locate and manipulate label information, they found it difficult to assess label claims. Szykman *et al.* (1997) claim that the lack of knowledge and scepticism towards ethical labels have a significant negative effect on sales.

Product attributes may be divided into search, experience and credence attributes (Caswell and Mojduszka, 1996; Loureiro *et al.*, 2002). Search attributes such as price, size and colour can be judged prior to the purchasing decision. Experience attributes such as product quality and taste can be assessed after using the product. Credence attributes can be judged neither before nor after buying the product. The ethical quality of a product is an example of a credence attribute. A credible ethical product label can convert this credence attribute into a search attribute (Caswell and Padberg, 1992; Caswell and Mojduszka, 1996; Pant and Sammer, 2003). Increasing the amount of certification proof can raise ethical label credibility. This can be effectuated by raising the credibility of the label issuer, i.e. the support of people or institutions with high public trust, and the perceived quality of the information provided in a credible and transparent process (Loureiro *et al.*, 2002; Nilsson *et al.*, 2004).

Pant and Sammer (2003) in their analysis of the characteristics of an efficient label to promote sustainable consumption in Switzerland conclude that issuer credibility was one of the important conditions, and also Nilsson *et al.* (2004) in their study of fifty-eight ecolabels stress the importance of reputable certification agents or endorsers. Two different types of issuer can be envisaged, i.e. governments and non-governmental organizations (NGOs). According to Nilsson *et al.* (2004) NGOs with broad stakeholder support are perceived to be most trustworthy. Both national and European governments could endorse ethical labels. In their study of social labels, Zadek *et al.* (1998) claim that, although European labels would be the most comprehensive and standardized approach, they would not work because they do not take differences in national identities, specific social movements and product types into account. On the other hand, in their study of animal welfare labels, Blokhuis *et al.* (2003) state that European labels for farming and animal products are necessary for better marketing. The label issuer as such seems to be of relatively high importance in the ethical buying decision, but the type of endorser also appears to be relevant.

Ethical labels should be easily understandable and contribute to less information cost, i.e. reduce the time it takes to find an ethical product. Therefore, some authors argue that attracting attention is important, not providing more information (Singh and Cole, 1993; Zadek *et al.*, 1998; Pant and Sammer, 2003). On the other hand, in a study on Fair Trade labels by Mielants *et al.* (2003) consumers complained about their limited knowledge of Fair Trade labels and their meaning, and a MAFF (2000) study in the UK revealed that around 50 per cent of consumers wanted more information on labelling and had difficulty finding the information they were looking for. In any case, the attractiveness of a Fair Trade label will to a certain extent depend on the perceived quality and quantity of information about the label. Too much information may be perceived as information overload which leads to confusion, or it may be perceived as an important criterion for the attractiveness of ethical products.

Pricing

In recent years, there has been evidence that the ethical consumer sees a more direct link between his or her buying behaviour and the ethical problem itself (Tallontire *et al.*, 2001). According to a study by Hines and Ames (2000), 51 per cent of the population have the feeling of being able to make a difference in a company's behaviour and 68 per cent claim to have purchased a product or a service because of a company's 'Corporate Social Responsibility' (CSR) reputation. On average, 46 per cent of European consumers also claim to be willing to pay substantially more for ethical products (MORI, 2000). However, the willingness to take an ethical label into account and/or to pay a certain price premium for it seems to depend on the type of label. According to one study, American consumers were prepared to pay a 6.6 per cent price premium for environmentally friendly products (The Roper Organization, 1990), while French consumers were willing to pay 10 to 25 per cent extra for apparel not made by children (CRC-Consommation, 1998), and Belgians were prepared to pay on average 10 per cent more for Fair Trade-labelled products (De Pelsmacker *et al.*, 2005a). Loureiro *et al.* (2002) concluded that the willingness to pay for an ecolabelled food product was an extra 5 per cent. Shaw and Clarke (1999) concluded that in the UK, Fair Trade was the most important ethical issue of concern. Maietta (2003) found that the consumers in his study were willing to pay a price premium of 9 per cent for Fair Trade coffee and 25 per cent for organic (biologically grown) coffee.

In Figure 8.1 the average price differences between Fair Trade and non-Fair Trade products offered in Belgian supermarkets are given. In the qualitative study by Mielants *et al.* (2003) and the survey by De Pelsmacker *et al.* (2005b), price emerged as one of the most compelling reasons for most people not to buy Fair Trade products. De Pelsmacker *et al.* (2005a) tried to assess which consumers were willing to pay the actual price premium of about 30 per cent for Fair Trade coffee. Approximately 10 per cent stated they would be prepared to pay this extra price. Half of them were 'Fair Trade lovers'. About 90 per cent of this group was prepared to pay a 10 per cent price premium, as well as almost two-thirds of the 'Fair Trade likers' group. On the basis of this result, one might conclude that the maximum penetration of Fair Trade products at the present price premium level is about 10 per cent. This could potentially increase to 35 per cent if the price premium was reduced to 10 per cent. However, at present most ethical brands and ethical-label products, including Fair Trade products, have market shares of less than 1 per cent (MacGillivray, 2000).

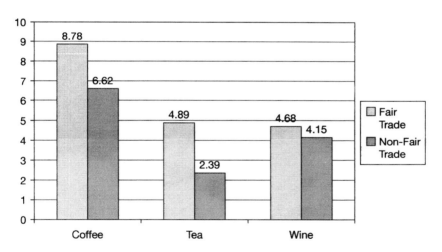

Figure 8.1 Average price in euro of Fair Trade and non-Fair Trade products (Belgium 2002)

Source: A. C. Nielsen, 2002

Distribution

In Belgium, mainly Oxfam and Max Havelaar market Fair Trade products. In the northern (Dutch-speaking) part of Belgium, Oxfam runs approximately 200 World Shops, operated by about 6,000 volunteers (only the fifteen largest shops have paid staff: thirty-five in total), focusing exclusively on food products. In the southern (French-speaking) part of Belgium, Oxfam has seventy-five

World Shops with 3,000 volunteers and seventy paid staff, focusing mainly on handicrafts. Oxfam also operates small sales outlets in schools, run by students and supported by teachers (Oxfam, 2003). In 2001, the Oxfam shops had a turnover of €5 million in food products (30 per cent of which is coffee) and €2.9 million in handicrafts (Krier, 2001). Oxfam not only sells its brand of Fair Trade products through its chain of World Shops, but also in ordinary supermarkets. Fair Trade products are also marketed by labelling organizations. Labelling organizations do not have their own brands or specialty stores, but issue their label to products of other manufacturers or retailers with which they have signed a contract which implies that these manufacturers or retailers will offer the Fair Trade labelled products in the non-specialized trade channels (ordinary supermarkets). In Belgium, the largest labelling organization is Max Havelaar. It is supported by a coalition of twenty-eight member organizations. It has contracts with fifteen licensees, mainly retailers who agree to sell part of their private label brands with the Max Havelaar Fair Trade label. Max Havelaar-labelled products are sold in more than 1,000 supermarkets in Belgium, and also in World Shops. They sell €5 million in retail value of mainly coffee and bananas, 55 per cent of which is outside the Fair Trade specialty trade channel (Oxfam, 2003).

In Table 8.2 the market shares of Fair Trade coffee in volume and in value in different types of regular supermarkets are given. Value shares are higher than volume shares as a result of the higher price of Fair Trade coffee. The Fair Trade share is marginal in small and medium shops and supermarkets, and slightly larger in supermarkets. The highest share is realized in hypermarkets. A marketing factor that has seldom been explored in the context of ethical buying behaviour is the effort it takes for consumers to locate and buy ethical products. Sometimes ethical products have their own 'shop in the shop', but equally often they are not even presented separately per product category. Often

Table 8.2 Market share of Fair Trade coffee in different types of shops (2001)

	Market share (kg) (%)	Market share (EUR) (%)
Total Belgium	0.5	0.6
Hypermarkets[a]	0.8	1.0
Supermarkets[b]	0.4	0.4
Medium supermarkets[c]	<0.1	<0.1
Small supermarkets and shops[d]	<0.1	<0.1

Source: A. C. Nielsen (2002).

Notes:
[a] For instance, Colruyt, Delhaize, Carrefour.
[b] For instance, AD Delhaize, GB Partner.
[c] For instance, Aldi and Lidl.
[d] Surface area of < 400m^2.

ethically labelled products are mixed with brands of the same product category or hidden away on supermarket shelves where they only get a very limited amount of shelf space. Therefore they are not so easy to find. Furthermore, some types of ethical products (such as Fair Trade products and, to a certain extent, bio-products) are primarily sold in specialty shops. Consumers who are increasingly pressed for time find it very inconvenient to have to buy their Fair Trade coffee or bio-vegetables (organic) in different shops than their other groceries. Mielants *et al.* (2003) found this factor to be one of the most important reasons not to buy Fair Trade products. In the De Pelsmacker *et al.* (2005c) study, the majority of respondents felt it was more convenient to be able to buy these products in ordinary supermarkets. Fifty-three per cent of respondents preferred the Fair Trade products to be placed together within the product category to which they belong (e.g. all Fair Trade coffees together on the coffee shelves). About 27 per cent preferred all Fair Trade products to be in a separate department or on a separate shelf ('shop in the shop'). A minority of 20 per cent preferred the current situation, i.e. the Fair Trade labelled coffee brand may be found close to the same brand without a Fair Trade label.

Promotion

Many authors who have carried out research in the area of ethical products marketing agree that ethical products would benefit from more awareness building and promotion (Zadek *et al.*, 1998; Wessels *et al.*, 1999; Nilsson *et al.*, 2004). For instance, Maignan and Ferrell (2004) claim that businesses will only benefit from their CSR efforts if they communicate intelligently. In a focus group study by Mielants *et al.* (2003) even committed Fair Trade buyers admitted that they needed regular promotional reminders to remain motivated to buy these products. In a survey of 615 Belgian consumers (2004) (De Pelsmacker *et al.*, 2005b), for half of the respondents the most important reason not to buy these products was that they felt they did not have enough information to be convinced (50 per cent). In this same study, the respondents were also asked how they would like to be informed about Fair Trade issues and products. In Table 8.3 the average scores on seven-point scales are shown for various promotion and information tools and channels. They are presented in order of importance. The high scores of every communication channel reflect the perception of lack of information. The scores do not differ greatly. However, consumers clearly preferred point-of-sale (POS) information (including product labels) and liked mass media advertising less. Besides POS indications, they preferred 'credible' and 'semi-commercial' sources, such as brochures and documentaries. It is also remarkable that the endorsement of the government in terms of funding a mass media awareness campaign was not particularly welcomed.

Table 8.3 Communication tools for Fair Trade information

Communication tool	Score[a]
Indication in the store	6.05
Use of product labels	5.74
Information brochure	5.61
TV documentary	5.54
Promotion at festivals/events	5.49
Media campaign by the government	5.40
Mass media advertising	5.34
Billboards and posters	5.16

Note: [a] 1 = very inappropriate, 7 = very appropriate.

The relative importance of Fair Trade marketing instruments

In order to assess the relative importance for the buying decision of ethical labels (among which are Fair Trade labels) and label issuers, branding, distribution, and information and promotion of ethical products, in June 2004 a web survey was set up (for more details, see De Pelsmacker *et al.*, 2005c). In order to approach the multi-attribute decision character of (ethical) consumption behaviour as closely as possible, an adaptive conjoint measurement was used to measure the preferences consumers express when facing a coffee-buying decision. In a conjoint analysis, consumers are asked to indicate their preference for products with varying characteristics. The most important advantage of this type of analysis is that, by simulating real marketplace situations, conjoint analysis realistically models day-to-day consumer decisions and has a reasonable ability to predict consumer behaviour, especially in a multi-attribute decision context (Green *et al.*, 2001). Consumers indicate their preferences by making trade-offs between different attributes of a product. These trade-offs can be decomposed into part-worth utilities and importance weights for each product attribute and each attribute level. In this way, the importance of different attributes or criteria in the consumer's evaluation of the product can be studied (Green *et al.*, 1978).

The survey was published and the link was e-mailed to 5,545 people, i.e. the total administrative and academic staff of Antwerp University (3,020) and Liège University (2,015), two mid-sized Belgian universities, one in the north (Dutch-speaking) and one in the south (French-speaking) part of the country. Furthermore, the total staff of the VUM (Vlaamse Uitgevers Maatschappij) (510) was also surveyed.[2] The composition of the sample is given with respect to a number of socio-demographic characteristics and coffee-drinking behaviour. Women and older people are slightly over-represented. As may be expected, due to the specific university context, better educated respondents were also over-represented compared to the total Belgian population (Table 8.4).

Table 8.4 Composition of the sample

Whole sample (N = 750)		%
Gender	Male	40.7
	Female	59.3
Language	Dutch	58.1
	French	41.9
Level of education[a]	LS+HS	12.3
	HE	17.6
	HE(U)	70.1
Age	<34 years	53.6
	35–54 years	37.9
	>55 years	8.5
Income level	−2500€	58.4
	+2500€	41.6
Coffee	< = 5 cups/day	82.5
	> = 6 cups/day	17.5

Source: De Pelsmacker *et al.* (2005b: 512–530).

Note: [a] LS+HS = lower and higher secondary school, HE = higher education (non-university), HE(U) = higher education (university). Income level is net monthly household income.

In order to be able to conduct a conjoint analysis, a specific product environment has to be defined that is kept constant during the study, as well as a set of product attributes (characteristics) and levels of these attributes that are relevant given the subject under study, and for which preferences will be measured. In this study, the respondents were told that they faced the task of expressing their preference for different types of mild coffee. Subsequently, they were exposed to various trade-off tasks in which they had to indicate their preferences on 7- or 9-category scales. The attributes and attribute levels involved in these trade-off tasks are given in Table 8.5.

Four types of ethical labels were defined as levels of the label attribute. A Fair Trade label was defined as a label guaranteeing a fair price to farmers in developing countries that enables them to invest in their own production methods and in that way plan their future. A social label was defined as a label to ensure that products were manufactured with respect for the rights of employees (no child labour, no forced labour, free labour unions). A bio-label guarantees that herbicides, pesticides, non-natural ingredients or genetically modified ingredients were not used. Finally, an ecolabel relates to environmentally friendly products, manufactured on the basis of production processes that use as little energy, water or dangerous products as possible, and that lead to products which are recyclable.

Table 8.5 Attributes and attribute levels used in conjoint analysis

Attribute	Level
Type of label	Fair Trade label Social label Bio-label Eco-label
Label issuer	The label is issued and controlled by the Belgian government The label is issued and controlled by the European government The label is issued and controlled by an independent non-governmental organization
Label information	The pack of coffee only has a label on the front of the pack The pack of coffee has a label on the front and explanatory information on the back of the pack
Distribution	The pack of coffee can be bought in a special department of a supermarket, together with other Fair Trade/Social/Bio- or Eco-products, distinct from the 'normal' coffees The pack of coffee is grouped with the other 'normal' coffees The pack of coffee can be bought in specialty stores (for instance, Oxfam World Shop)
Promotion	The label is advertised on radio and television and in newspapers and magazines An informative brochure is mailed to you
Branding	Manufacturer brand (examples of well-known Belgian manufacturer brands of coffee were given) Store brand (private label) (examples of well-known Belgian store brands of coffee were given)

Source: De Pelsmacker *et al.* (2005c: 512–530).

The attributes 'distribution', 'promotion' and 'branding' measure three marketing aspects. The attribute levels defined reflect the debate on the appropriate distribution and promotion tactics discussed in previous sections. The sixth attribute, type of brand, assesses the importance of a traditionally strong attribute (the brand) of fast-moving consumer goods (such as coffee), and the extent to which a manufacturer brand or a private label is an important factor when forming preferences towards ethically labelled products.

The 'label issuer' attribute contains three levels: the local government, the European government, and a non-governmental organization. The 'label information' attribute contains two levels: the mere mention of the label on the front of the pack and additional product information on the back of the pack.

Table 8.6 Relative importance of attributes and part-worth utilities of attribute levels – full sample (2004)

	Relative importance of attributes, part-worth utilities
Number of respondents	750
Type of label	*19.82%*
Fair Trade[1]	23.39[2,3,4]
Social[2]	8.55[1,3,4]
Bio[3]	−20.75[1,2,4]
Eco[4]	−11.19[1,2,3]
Issuer of label	*17.48%*
Belgian government[1]	−6.85[2,3]
European government[2]	3.63[1]
Non-governmental organization[3]	3.22[1]
Label information	*15.00%*
Label only	−41.90[2]
Label + extra information	41.90[1]
Distribution	*24.54%*
Distinct department[1]	−1.47[2,3]
Same department[2]	45.60[1,3]
Specialty store[3]	−44.13[1,2]
Promotion	*11.37%*
Mass-media advertising	12.45[2]
Informative brochure	−12.45[1]
Brand	*11.79%*
Manufacturer brand	16.74[2]
Store brand	−16.74[1]

Source: De Pelsmacker *et al.* (2005c: 512–530).

Note: Percentages in rows indicate the relative importance of attributes. Superscripts indicate the part-worth utilities of the attribute levels that are significantly different from each other at the 5 per cent level. For instance: a Fair Trade label (23.39) is significantly more preferred than the other three label types, hence the superscripts 2, 3 and 4.

Table 8.6 summarizes the relative importance of each attribute and the part-worth utilities of each attribute level for the full sample, as well as the significance of the differences between part-worth utility levels. Paired sample t-tests indicate that all average importance differences between each pair of attributes are strongly significant, except for the difference between label information and brand, which is not significant. The distribution attribute generates the highest average importance (24.54 per cent), followed by type of label (19.82 per cent), issuer of label (17.48 per cent), label information (15.00 per cent), brand (11.79 per cent) and promotion (11.37 per cent).

Within each attribute, all part-worth utilities are significantly different from each other (t-tests). Within the distribution attribute, respondents assign the largest utility to a situation where the pack of coffee is grouped with the other 'normal' coffees. This attribute level scores significantly better than a situation where the pack of coffee may be bought in a special Fair Trade/social/bio/eco store department and much better than a specialty store. The second most important attribute is the type of label. Within this attribute a Fair Trade label generates the highest utility. The second highest utility is attributed to the social label, which in turn outperforms the ecolabel and the bio-label. The reported utility scores are all significantly different from each other. Of third importance is the issuer of the label. When the issuer is the Belgian government, a significantly lower utility is attributed to this level as compared to the European government and a NGO. The utilities of the latter two do not differ significantly. The label information attribute is of fourth importance. Extra information on the back of the pack besides a front label generates a significantly higher utility compared to only a front label. With regard to the brand attribute, a manufacturer brand generates a significantly higher utility than a store brand. Finally, for the promotion attribute, it appears that mass-media advertising generates a significantly higher utility than an informative brochure sent to the consumer.[3]

Taken together, the most attractive product is a manufacturer brand of coffee with a Fair Trade label with extra information on the pack, issued by the European government, distributed in ordinary supermarkets, displayed on the coffee shelves and mass-media advertised. The least attractive product is a store brand of coffee with a bio-label, issued by the Belgian government, distributed in specialty stores and promoted by means of an informative brochure.

The first remarkable result of the study is that social labels and especially Fair Trade labels are significantly more preferred than ecolabels and bio-labels. To a certain extent, this result may be qualified as surprising. One could have expected that biological and environmentally friendly production were of more concern to people, because they could be expected to be more involved with the direct environment in which they live. Social and Fair Trade production are issues that are often relevant for people in far-away countries. Perhaps one of the reasons for the current result is that ecological and biological products and production methods have been important issues for many years now (at least in Belgium), and therefore people find them less appealing, while Fair Trade products are relatively new and have been subject to less public scrutiny. In any case, the products with social labels are very rare in Belgium, but this does not prevent consumers from preferring them the most, in principle. Perhaps this result is also due to the bias towards more highly educated consumers in the sample.

The issuer of the label is the third most important attribute after the type of label and the distribution method. As such, the issuer is more important than the brand, the amount of information and the promotion strategy. Belgium is an internationally oriented country in which the European institutions are very visibly represented. Furthermore, due to the historical differences between the Dutch- and French-speaking parts of the country, the Belgian identity is not very well developed. This may account for the lower preference for labels that are endorsed by the national government and the higher credibility of non-governmental or European endorsement.

One of the most remarkable, though not surprising, findings of this study is the relative importance of the availability of ethical products in the distribution channel. Although this aspect received little attention in previous research, it is by far the most important attribute in this study. Furthermore, consumers have a clear preference for a distribution strategy in which ethical products are put on the same shelves as non-ethical products, but as a separate group. The way in which ethical products are promoted outside the shop by means of mass (marketing) communications techniques appears to be of relatively little importance.

Finally, the type of brand on which an ethical label is placed is of relatively minor importance. This is somewhat surprising, but may be due to the nature of the product studied. Belgium is a country in which store brands of coffee traditionally have large market shares. Many people are used to drinking store brand coffee, and apparently do not perceive it to be of lower quality than manufacturer brand coffee. Furthermore, the apparent focus of this study on ethical labels (Fair Trade, social, bio, eco) may have distracted the attention of respondents away from 'obvious' attributes such as the brand, and may therefore have underestimated its relative importance in this study.

Conclusions and implications

Belgian consumers seem to have a rather good knowledge of the Fair Trade concept, and their attitude towards Fair Trade in general is quite positive. However, most consumers find the Fair Trade products too expensive, and too much effort has to go into buying them. Although the personal service in specialty shops is evaluated positively, the shops are perceived as too sober and too 'alternative'. The most important reason for respondents to buy Fair Trade products is because it gives a fair price to the farmers and manufacturers of developing countries. Besides the high price and the poor distribution, another important reason to keep the respondents from buying Fair Trade products is lack of good information. More general information on the Fair Trade concept and more package information is welcomed, leading to more credible labels and brands. Especially in-store information is preferred, although each communication medium is considered to be an appropriate promotion channel for Fair Trade products.

This investigation into the relative importance of different characteristics of ethical products and marketing practices of ethical products shows that consumers attach most importance to the distribution strategy of Fair Trade products. Fair Trade products should be available in ordinary supermarkets and be presented along with non-ethical variants of the same product category. In order to broaden the appeal to other consumers, ethical products also have to be available in non-specialized distribution channels. Fair Trade-labelled products are by far the most preferred, followed by social label products. Eco- and bio-labels are less appreciated. European government labels, or labels issued by non-governmental organizations are most preferred, as opposed to national (Belgian) government-endorsed labels. Consumers prefer extra information on the package, in addition to a label. Out-of-shop promotion of the label and the type of brand are of minor importance. Consumers show a slight preference for mass-media advertising (as opposed to an informative brochure) and for manufacturer brands, as opposed to store brands.

Highly educated respondents, older people and people with a higher income are less sceptical and less indifferent, more inclined to action, and are more readily prepared to pay the Fair Trade price premium. Especially older and less educated consumers attach a lot of importance to the label issuer.

The results have a number of implications for the marketing strategies and tactics of NGOs endorsing Fair Trade products by means of labels, and for organizations, manufacturers and retailers marketing Fair Trade products:

- Older, more highly educated and higher-income consumers appear to be the obvious socio-demographic target group for Fair Trade products. Traditionally, the marketing focus of Fair Trade organizations (in Belgium) is on children (in schools) and middle-aged people. It may be worthwhile to consider special marketing efforts towards the older age groups. The strategy to appeal to customer groups for which idealism is an important value seems to be the most appropriate one.
- As such, the knowledge of and attitudes towards Fair Trade are relatively positive. Therefore, putting more effort into awareness-raising campaigns does not seem to be called for. Consumers understand the unique selling proposition of Fair Trade products very well, and a lot of them seem to be prepared to endorse them. The problem lies in the 'attitude–behaviour gap': positive attitudes are in many cases not translated into buying behaviour.
- Fair Trade labels are by far the most attractive ones to Belgian consumers. However, the credibility of the label issuer is very important. Products with Fair Trade labels issued by European governments and NGOs are more attractive than nationally endorsed labels, and therefore stand a greater chance of being taken seriously and – consequently – of being purchased. The Belgian government should therefore refrain from adding to the plethora of labels and to consumer confusion, but instead label initiatives should be standardized and integrated on a (non-governmental) pan-European level. This implies reducing the number of national and perhaps also product-specific labels to a smaller set of more powerful and credible labels.
- For the vast majority of consumers, the price premium of Fair Trade

products is a significant hurdle that impairs a higher penetration of Fair Trade products. If Fair Trade organizations are incapable of lowering the price, the government may consider shifting the funds that are now used to increase the awareness of the Fair Trade issue to lowering their price, for instance, by means of tax measures that favour Fair Trade products or penalize non-Fair Trade products.

- Offering Fair Trade products in specialty shops is not enough. Fair Trade products should be much more visibly available in regular supermarkets. Label issuers should also try to convince manufacturers and retailers to put the ethical label products on their shelves, as a visible subgroup within product categories.
- Consumers state that they would buy more Fair Trade products if more and better information about Fair Trade were available. This information should be primarily available at the point of sale and on the product; for instance, by providing additional information on the package.
- Out-of-store promotion, though perceived as less important, should take the form of mass-media advertising rather than of low-profile direct communication.
- Manufacturers and distributors of labelled products should realize that ethical labels are more attractive when put on manufacturer brands rather than store brands. Perhaps a genuine Fair Trade brand would be the most effective marketing strategy to combine both the ethical quality of a product and the impact of a strong brand.

Notes

1   In Belgium, a number of Fair Trade organizations are active (Krier, 2001): Oxfam World Shops (Wereldwinkels and Magasins du Monde), Fair Trade Organisatie (a subsidiary of the Dutch FTO), Maya Fair Trading, an organization focusing on honey and honey-based products, and Max Havelaar Belgium, a Fair Trade label organization accrediting coffee and bananas in more than 1,000 supermarkets. Sales of Fair Trade products in Belgium are only marginal for sugar and chocolate products, slightly more meaningful for tea and wine, and relatively substantial for coffee. Max Havelaar estimates that the Fair Trade coffee market in Belgium was 582,203 kg in 2001; 640,482 kg in 2002 and 762,000 kg in 2003. Total coffee sales were estimated at 35,794,182 kg in 2003, which is equivalent to a market share of Fair Trade coffee of 1.6 per cent in terms of volume. Direct sales of coffee to companies and governmental institutions are not included in the previous total sales figures. Including these sales would add up the total to 52,778,000 kg in 2001, which leads to a market share of Fair Trade coffee of 1.1 per cent in terms of volume. In a survey of 615 Belgian consumers (De Pelsmacker *et al.*, 2005c) about 60 per cent of the respondents indicated that they purchased Fair Trade products less than five times per year, while about 10 per cent claimed that they purchased Fair Trade products more than twenty times per year. More than 60 per cent estimated that they spent less than 60 euros per year on Fair Trade products, and 14 per cent claimed to spend more than 100 euros per year.

2   A total of 750 completed questionnaires were received: 402 employees from Antwerp University (response rate: 13.31 per cent), 34 employees of VUM (response rate: 6.67 per cent), and 314 employees from Liège University (response rate: 15.58 per cent).

3   When the connotation 'significant' is used, this refers to the p-value of t-tests < 0.05.

9 Consumption as a solidarity-based commitment

The case of Oxfam Worldshops' customers

Gautier Pirotte

Fair Trade and the responsible consumers

For several years now, the movement in favour of more equitable international trade has been gaining public interest. Surveys (see Box 9.1) seem to show that the general public is better and better informed about the principles of Fair Trade, as well as about its products and its promoters.

Box 9.1 Commitment to Fair Trade: Fair Trade policy

According to the criteria defined by the European Fair Trade Association (EFTA, 1998: 5–6), the movement for more equitable commerce relies upon certain principles. First of all, Fair Trade implies the 'fair remuneration' of small-scale producers of the Southern hemisphere, meaning remuneration for production that allows producers and their families to attain a sufficient standard of living. This especially implies the establishment of a 'fair price', of advanced payments if necessary, and of long-term commercial relationships. Furthermore, it can be partially classified within the issue of sustainable development, as it proposes, for example, a sustainable method of production from an ecological and economical standpoint; in other words, a level of production that meets the needs of current generations without compromising the capacity of future generations to satisfy their own needs. Fair Trade also aims to reduce the commercial channel to its principal protagonists by eliminating unnecessary speculators and intermediaries. It claims to strive for preferential treatment in production, trade and sales implicating local production by small and medium-sized businesses based on fair distribution of income. In the North, it develops awareness campaigns about the relationships between consumers' choices and the living and working conditions of producers in the South as well as awareness campaigns aiming to change unfair structures of international trade.

Somewhat along the same lines as the organic food movement, the main actors involved in distribution and even production and import seem to be keeping a close eye on the advancement of this movement. At the same time, the Fair Trade movement, decades old already, seems to be experiencing a renewal and integrates itself into the constellation of alter-globalization movements, as much by its principles as by the actors that embody it.[1]

Although it has been the subject of numerous studies, this old project which aims to modify commercial relationships between the Northern and Southern hemispheres of the planet to make them more balanced (and some would say more ethical) has aroused recent interest from the scientific world. In order to demonstrate the validity of these initiatives, impact studies have been conducted by experts (see e.g. OPM/SMG, 2000; Hopkins, 2000) sponsored by the organizations at the root of these Fair Trade projects (like Oxfam Fair Trade or Max Havelaar). At the present time, it is no longer simply a question of measuring the impact of these Fair Trade initiatives on the licensed beneficiaries, but of understanding the dynamics specific to Fair Trade (see also Le Velly, Chapter 14, this volume) in order to replace the initiatives of alternative commercialization at the heart of a social or cultural change process in the Southern hemisphere (Charlier *et al.*, 2000). The studies on equitable consumerism, beyond the surveys that aim to measure the level of popularity of this movement and its products, are nevertheless quite rare. The sociologically oriented surveys included in the works of research teams from the universities of Liège and Antwerp, coordinated by Marc Poncelet (2005) for the Belgian Science Office, are quite innovative. The main teachings of these surveys are presented here. The consumption of Fair Trade products is interpreted as the expression of a commitment in favour of someone in a far-away place. In this perspective, the analysis of Fair Trade consumption is not only worthy of a classical analysis of the act of consuming (e.g. in terms of satisfaction of needs, of the logic of distinction, of the expression of habits of class) but implies a more socio-political approach, linking consumption to a particular engagement.

This survey on the engagement of the consumer in Fair Trade appears to be all the more relevant in that the movement's capacity for change in favour of Fair Trade rests upon the hypothetical capacity of Northern consumers to produce change in the South by their choices. Fair Trade, while partially distinguishing itself by the direct North–South dimension of the issue and a more explicit integration in the theme of sustainable development, is very closely related to other movements that rely on the mobilization of consumers (ethical commerce, boycotting of genetically modified organisms or the consumption of organic products) by frequently calling upon consumer power to introduce a change in the ecological, economic and/or social production conditions. We therefore refer to the 'ethical consumer' (Harrison *et al.*, 2005), the 'consomm'acteur' ('consumer-actor') in French (Gendron *et al.*, 2003), or the 'responsible consumer' (see also Ruwet, Chapter 10, this volume).[2]

Our Western societies are largely described as over-consuming ones, where personal fulfilment, the maximizing of the potentials of our bodies and relationships, unlimited health and the conquering of personal time and space are dominant (Lipovetsky, 2003: 74). But some approaches see the consumer as able to cease to be this individualist in search of the selfish satisfaction of his needs, to become the pillar of democratic contestation in the era of globalization.[3] Buying a Fair Trade product, according to Serge Latouche:

> is a way of affirming political mediation in commercial exchange, and therefore also in the solidarity with distant and unknown partners, without disregarding their existence or proving indifferent to their fate.
>
> (Latouche, 2003: 258, trans. author; see also Latouche, Chapter 12, this volume)

The study mentioned above, in which our team has been involved with other scholars (see also De Pelsmacker *et al.*, Chapter 8, this volume) aimed to test the prevalence of this type of consumers' rationale, using a study conducted on the clientele of shops that sell Fair Trade products (craft production and food) in Belgium: the Oxfam-Magasins du Monde and Oxfam-Wereldwinkels (both referred to in this chapter as Oxfam World Shops).

The social commitment as the forgotten dimension

When reflecting in the North on the consumption of Fair Trade products, one will notice the remarkable increase in sales during recent years (about 20 per cent a year since 2001). The aggregate net retail value of Fair Trade products (labelled or not) sold in Europe in 2005 exceeds 660 million euros, representing over twice the figures of five years ago (Krier, 2005). In view of this optimistic trend, one may be surprised by the fact that, despite the efforts made by Fair Trade organizations, the share of these products in the global world trade remains very small. Nowadays, the sale of Fair Trade products is no longer limited to specialized stores and fair consumption is no longer only an act of (gastronomic) courage or Christian charity. However, although it is not a covert act, the consumption of Fair Trade products is not always done clearly, as the main players of the movement can confirm. In 2002, Fair Trade did not even account for 0.01 per cent of world trade (0.008 per cent), i.e. 350 million euros for Fair Trade compared to 4,000 billion euros for international trade (Lecomte, 2003: 21).

One notes that efforts have been made within the Fair Trade movement to get the products out of their traditional distribution networks (causing questions of conscience for some of the most committed members). While few Fair Trade products are sold by mail order or on the Internet (our work specifically highlights this), Fair Trade products have a distribution network that is equivalent, in the number of points of distribution, to that of a multinational petroleum company (according to a study made in Europe[4]). In Belgium, Fair Trade products are sold through 227 Oxfam World Shops (160 in Flanders and 67 in the French-speaking community) and several hundred supermarkets belonging to the eight largest chains.

While sales and consumption of Fair Trade products are not very significant, this is not due to a lack of consumer interest. Surveys regularly show that there is a significant increase in awareness of the movement, its principles, its participants and the products. In a pan-European survey conducted by the European Fair Trade Association (EFTA) in 1998, about 62 per cent of Belgians were aware of the existence of Fair Trade products but only 17 per cent said they were ready to purchase them. The telephone survey conducted by IDEA Consult in June 2002, involving 1,005 Belgian consumers, showed that 52 per cent of those questioned did not know the Fair Trade principles while 15 per cent were actively aware[5] and 33 per cent were passively aware.

In France, surveys show very similar results but with a few differences. Tristan Lecomte, when studying recent opinion data,[6] identified a significant tendency: a major rise in awareness of Fair Trade. The surveys show that there is a large increase in Fair Trade's prompted recognition between 2000 and 2003.[7] In October 2000, 9 per cent of respondents knew about Fair Trade. In May 2002, this had increased to 24 per cent. Recognition of Fair Trade falls with age. For people under 60, the rate is around 35 per cent (over 36 per cent among under-25-year-olds). For people over 60, it drops to 25 per cent. In the October 2002 IPSOS survey, the rate of recognition rose to 32 per cent. More importantly, the recognition is improving. In October 2001, the largest group of respondents (41 per cent) associated Fair Trade with the campaign against child labour. Two years later, a new consensus (39 per cent) associated Fair Trade with the issue of a new commercial balance between North and South. Thirty-one per cent of respondents thought that Fair Trade was a way of dealing with poverty in the South. These opinion surveys also clearly demonstrate the difference between the buying intentions for Fair Trade products that are always very high – an indicator of the social attractiveness of the movement – and actual purchases, which, as we have emphasized above, are weak.

Multiple factors can be proposed as explanations for the size of the Fair Trade market share at the consumer level (and also at the production level). These factors may include the identification of products, the consequences of purchasing Fair Trade products, the price, the flavour, the packaging, and the perception or not of the labels. These are the questions usually investigated in Fair Trade market research (see De Pelsmacker *et al.*, Chapter 8, this volume). The overall sociological hypothesis that we want to test involves presenting fair consumption principally as an action that demonstrates a commitment to a cause: greater justice and solidarity in international trade. For Fair Trade consumption to be considered as a commitment there has to be an assumption that the consumer has specific resources and competences that are not equally shared among all the social players.

Sociological analysis of commitment shows that interest and commitment for 'someone far away' requires that a person has various resources, with knowledge and experience of the South being the most important. However, few people have in-depth knowledge and experience of living in a developing country, factors which could prove fundamental to understanding Fair Trade product purchases. How else can we explain that more than 80 per cent of employees and volunteers in Belgian cooperation and development non-governmental organizations (NGOs) consume these products (Stangherlin, 2004)?

This analysis may account for the small share of market held by Fair Trade in the Western world and the difference between awareness of the movement and purchasing behaviour. Many factors can explain this gap between the buying intentions and the final purchase of Fair Trade products. They are, most of the time, studied by marketing approach (see De Pelsmacker *et al.*, Chapter 8, this volume). But there is so far a lack of interest in studying Fair Trade consumption as a socio-political act. We take as a starting point for our reflection the social movement dimension of this trend for a North–South trade more ethic and fairer. For example, we may wonder if purchasing Fair Trade products is not easier when consumers can mobilize some specific resource (e.g. education, time, visions of North–South relations, specific knowledge about developing countries).

Despite the current widespread availability of Fair Trade products, it is possible that the purchase of such products remains an indirect action because, in addition to issues related to their characteristics (price, quality, packaging, distribution), fair consumption is a way of showing a commitment to a specific cause. Fair consumption could imply that one has first acquired specific resources that result in a tendency to act in accordance with the movement's principles. This may provide an explanation for the fact that greater Fair Trade product availability has not resulted in an explosion in demand for these products; it may be due to a specific social reality in which purchasing Fair Trade products would remain reserved to a specific type of client. We need to define this clientele in greater detail so as to understand its specificities and the potential obstacles and facilitators for increasing its numbers. We will use three levels to study this clientele: sociological profile (level 1); commitment to Fair Trade (level 2); and tendency to commitment (level 3).

Samples

Questionnaires were sent by mail to a representative sample of 5,000 Belgians who are responsible for the daily purchasing in their households; 1,200 of the same questionnaires were distributed among consumers of forty Oxfam World Shops. In addition, thirty questionnaires were sent to twenty Oxfam World Shops in the Dutch-speaking part of Belgium, and to twenty Oxfam World Shops in the French-speaking part. A total of 1,138 questionnaires were received: 799 of the Belgian sample (response rate of 16 per cent) and 339 of the World Shop sample (response rate of 28 per cent). After cleaning up the data file (e.g. deleting respondents who were not responsible for the purchases in their household and deleting the data of respondents who did not fill in 10 per cent or more of the scale-questions) there were 615 respondents for the Belgian sample and 243 for the World Shop sample.

Table 9.1 Commitment to Fair Trade: summary table of the central hypotheses

Level of analysis	Hypotheses	Indicators
Sociological profile	1 Oxfam clients have sociological characteristics different from those of other Belgian consumers.	Sex; age; education level; socio-professional status and employment; education and religious practices; region/language; number of children and age of youngest; marital status; income; political position.
Commitment to Fair Trade	2 Oxfam clients are not simply different in their sociological profile but also by a more committed relationship with Fair Trade.	*Means of access* to Fair Trade; *Reasons for buying* Fair Trade products; *Frequency of purchasing*; *Budget used* for buying Fair Trade products.
Type of commitment	3 This profile difference also involves different types of commitments, specifically in favour of the South, based on different perceptions of the causes of underdevelopment.	Involvement in the associative network; Types of 'committed' actions; Perception of the problems in developing countries favouring a commitment for the local populations (causal attribution).

We would like to remind the reader of an essential element for understanding the data that will be presented hereinafter. We approached the two populations being investigated in two different ways: one was a representative sample based on quotas, while the other was more random and involved respondents who had in common the fact that they had visited one of the selected Oxfam World Shops during the study period. Neither of these methods is without bias. The low rate of replies within the 'Belgian consumer' sample has without doubt created significant bias (see below). It should not be forgotten either that among the sample of Belgian consumers there may also be some regular Oxfam clients. This significant factor limits the comparison options.

Results

First level of analysis: the sociological profiles

While the preliminary surveys (EFTA [1998] in Belgium; Lecomte [2003] in France) usually suggested that a 'fair consumer' is generally a highly educated female, our work has not shown significant correlation between gender or education and being an Oxfam client. Furthermore, there is no correlation between education and religious practices and being an Oxfam client, which could show – at the level of the clients at least – the distance between the movement and its initial religious starting point (Roozen and Vanderhoff, 2002). On the other hand, our survey showed a number of factors of influence between being an Oxfam client and some of the variables. Compared to the sample of Belgian consumers, young people (between 18 and 24) and older age groups (from 55 to 75 and over) are over-represented[8] among Oxfam World Shops clients. As regards the number of children, two groups are over-represented among the Oxfam clients none (24.31 per cent of Oxfam sample and 17.88 per cent of Belgian consumers sample) and more than two children (25.89 per cent of Oxfam sample and 21.52 per cent of Belgian consumers sample). In the Oxfam sample, students (3.6 per cent compared to 0.3 per cent in the Belgian consumers sample) and retired people (16 per cent compared to 9.9 per cent)[9] are over-represented among Oxfam clients. In terms of profession (past or present), the differences between the two samples[10] are weakly significant with perhaps the exception of salaried staff (administrative or non-administrative roles)[11] that are over-represented and specialist or qualified workers that are under-represented among Oxfam clients.[12] Divorced, widowed and single people[13] (particularly) are over-represented among Oxfam clients. For example, 7.2 per cent of respondents are single in the Oxfam sample compared to 1.72 per cent of Belgian consumers.[14] Finally, we note that low incomes (less than 1,000 euros per month) and high incomes (more than 5,000 euros per month) are over-represented among Oxfam clients.[15] These results[16] lead to two profiles of people who go most often to Oxfam World Shops (Table 9.2).

This does not mean that the 25 to 34 and 35 to 44 age groups are not part of the Oxfam clientele but, compared to an average sample of Belgian consumers, they are seen less frequently in Oxfam World Shops. This seems to

Table 9.2 Two profiles of Oxfam customers

'Young' profile	'Over 45' profile
Young (18–24 years)	> 45 years
Student	Retired people
No children	> 2 children (> 25 years)
Low income (< 1,000€)	High income (> 2,500€)
Single	Widow/widower or divorced

be consistent with personal resources commitments and uses, especially time, and partially education and money (for the elderly). This might provide us with an explanation for the under-representation of the group of people between 25 and 44 in terms of 'biographic availability' (Stangherlin, 2004): it is a time of their lives when their active life is beginning, they are creating a home and a family and consolidating their career choices. These are biographical events that tend to limit the mobilization capacities of individuals.

Second level of analysis: different commitments to Fair Trade

In terms of purchases, World Shops clients are, as expected, more frequent users of Fair Trade products: 42.9 per cent of them have purchased Fair Trade products more than twenty times during the previous twelve months compared to 9.8 per cent of Belgian consumers. Nevertheless, nearly 9 per cent of Oxfam clients had purchased no Fair Trade products within the past year. Oxfam clients are not only more regular buyers of Fair Trade products but they also spend more. More than half the sample of Oxfam clients had spent over 101 euros (101–250 euros and >250 euros) on Fair Trade products during the previous year compared to 13.4 per cent of Belgian consumers.

In terms of Fair Trade product availability, both samples gave Oxfam World Shops as the main access channel; 47.6 per cent of Oxfam clients discovered Fair Trade through Oxfam World Shops compared to only 8.6 per cent through supermarkets. These shops are also judged as the main access channel to Fair Trade products for Belgian consumers (24.6 per cent) but only just ahead of the supermarkets (21.9 per cent).

These figures seem to confirm the idea that purchasing in an Oxfam World Shop goes beyond a simple consumption action. At a time when the number of outlets for Fair Trade products is increasing, nearly one in two Oxfam clients claims to have first become aware of these products in these shops. There is a considerable difference within this group between this means of access and others. The press (8.14 per cent) and publicity campaigns (5.6 per cent) do not score very well whereas, among the Belgian consumers, these channels are the third largest (12.4 per cent and 8 per cent, respectively) before friends (9 per cent), the family (8.5 per cent) and schools (6.2 per cent).

The questions about what motivates purchases also show that Oxfam clients are more acutely aware of the consequences of their purchases. Responses from Oxfam clients may be divided into three main reasons: payment of a fair price to the producer in the South (75.8 per cent), respect for working conditions (64.4 per cent) and the dignity and self-sufficiency of the peasants in the South (62.5 per cent). These three items correspond fully with the messages that Fair Trade organizations (and especially Oxfam) try to get across to consumers.[17] This confirms the results of the market survey that showed a high level of understanding of Fair Trade by Oxfam clients (see De Pelsmacker *et al.*, Chapter 8, this volume). Conversely, replies from Belgian consumers are spread across a greater number of subjects even though the above-mentioned categories also dominate their responses (61.5 per cent, 56.2 per cent and 50.4 per cent, respectively). The Belgian consumers are more interested than the Oxfam clients in the product characteristics without any specific link to the Fair Trade movement. As a result, the more important reasons for purchasing for Belgian consumers are product quality (19 per cent compared to 9.8 per cent), environmental protection (24.7 per cent compared to 17.4 per cent),[18] the product's flavour (15.4 per cent compared to 8.6 per cent) and health aspects of the product (15.1 per cent compared to 5.8 per cent). The association of Fair Trade products with a cause is not absent among Belgian consumers. They give a higher result in terms of purchasing Fair Trade products for a 'good cause' (23.4 per cent compared to 14.8 per cent). They are also more likely than Oxfam clients to purchase the products out of simple curiosity (14.6 per cent compared to 5.6 per cent).

One can however presume that Oxfam clients consider the consumption as a political commitment. This is suggested by the over-representation of the 'Fair Trade', 'working conditions' and 'producer's dignity' replies for this group and also by the over-representation of 'bring about political and social change' (38.7 per cent) and the 'alternative consumption' aspect (10.5 per cent) that were less apparent among Belgian consumers (15.7 per cent and 5.6 per cent, respectively). One should also highlight the importance of 'World Shops' for their role in informing consumers and increasing their awareness. This may be seen through a question where we asked Oxfam clients to comment on the advantages of different distribution channels. The Oxfam clients emphasized the quality of the welcome in the shops (32.4 per cent compared to 2.73 per cent for supermarkets) but also mentioned the information available in the shops (14.8 per cent compared to 1.95 per cent in supermarkets).

Third level of analysis: the types of commitment

We wanted to test the theory that Oxfam clients are more committed citizens than ordinary consumers. For this, we wanted to evaluate if they had a greater degree of involvement in associations, were more active in public actions (petitions, demonstrations, gifts, ethical investments) and more oriented towards the issues of the South. Finally, as a recent survey (IDEA Consult, 2002) showed, people with an unprompted awareness of Fair Trade were more attentive to the problems of the South; we linked this commitment to a different way of perceiving the problems in these countries.

First, we asked respondents if they were members of any association and, if yes, which one(s).[19] There is greater non-involvement in associations (Chi^2 = 0.000) among Belgian consumers (35.2 per cent) than among Oxfam clients (22.3 per cent). This leads us to assume that Oxfam clients are more involved in the associative sector than are normal consumers. The types of associations most often mentioned by Oxfam clients were cooperation and development associations (24.2 per cent, Chi^2 = 0.000), cultural and artistic associations (23.1 per cent, Chi^2 = 0.000) and environmental associations (16.4 per cent). There was an over-representation among Oxfam clients of environmental associations (16.4 per cent compared to 8.2 per cent, Chi^2 = 0.000), pacifist movements (5.9 per cent compared to 1.1 per cent) and human rights associations (7.8 per cent compared to 3.8 per cent, Chi^2 = 0.012).

As they seem to be more involved in associations, are Oxfam clients also more involved in civic actions or, in other words, in activities during which they publicly display their commitment to any cause? One notes that there is an over-representation of Oxfam clients in the response 'product or service purchased because it is linked to a good cause' or 'support of an activity for a good cause' (very broad answer). Some replies may be considered as an extension to those given earlier. As a result, the ecologic element of activism (buy organic products: 58.6 per cent, support ecologist organizations: 36.9 per cent) indicates a greater level of participation by Oxfam clients in ecological organizations; the same applies to cooperation and development activities (awareness of problems of developing countries: 20.7 per cent, voluntary work for a NGO: 34 per cent). We also observed that the Oxfam client is more 'restless' than the Belgian consumer: while 12.4 per cent of Belgian consumers have taken part in a protest during the previous twelve months, this figure climbs to 30.1 per cent among Oxfam clients. Compared to civic actions of Belgian consumers, Oxfam clients are more likely to give financial support (financial donations, ethical investments, purchase products, foster children) than giving in kind (e.g. gifts of clothes). It appears that Oxfam clients are also more oriented towards relations with others based on justice (protests) and solidarity (financial support) than on aid (gifts of clothes), the traditional form of relationship with the South.

The analysis may be fine-tuned by using a typology of the kinds of solidarity relationships with 'far-away others', which was initially described by Kersting (1998) and tested empirically by Stangherlin (2004) as part of a thesis on militants in development NGOs in Belgium (Table 9.3).

Table 9.3 A typology of solidarity relationships

Solidarity-based relationship	Principles	Empirical translations
Aid	Universal standards. Context of another's suffering. Unilateral. No response from the person helped.	Charity. Gifts of clothes, food, medicine. Fostering children.
Solidarity	Specific standards. Exchanges between equals. Reciprocity.	Financial support (loan, gift distribution/counter-gift). Development projects.
Justice	Universal standards. Defence of inalienable rights.	Manifestation. Petitions. Lobbying. 'Political action'.

Source: Adapted from Kersting (1998) and Stangherlin (2004).

If we group some items (cf. Table 9.4.) using this typology, we observe that Oxfam clients and Belgian consumers participate, in terms of frequency, most often in a commitment that takes the form of charitable aid, but Oxfam clients develop this type of activity less than Belgian consumers. They are – sometimes – very greatly over-represented in the other forms of solidarity-based commitments, preferring more political or more egalitarian actions. One also observes that for a given type of activity, Oxfam clients prefer actions targeted at developing countries. This is most evident in financial support (48.1 per cent for developing countries and 29.7 per cent for the disadvantaged in Belgium) and to a lesser degree in their gifts of clothing, food and medicines (57.8 per cent for the South and 56.6 per cent for Belgium).

To explain the commitment to Fair Trade, we believe that we must analyse the perceptions of the socio-economic and political realities in the developing countries, since the most disadvantaged fringes of their populations (especially small producers) are usually the targets of Fair Trade actions.[20] We looked at how the problems of developing countries are perceived by the two consumer groups. We asked them if they agreed with a number of statements about the causes of the underdevelopment of these countries. We then identified two main categories of explanations: those related to internal factors from within the societies of the South (including political and economic corruption in developing countries; the mentality of the people; climatic conditions; demographic conditions) and those related to underdevelopment for external reasons (developing country debt and dependence on the North; unequal economic exchanges that are unfavourable to developing countries; the consequences of

Table 9.4 Social commitment of Oxfam customers and Belgian consumers

	Items	Oxfam customers (%)	Belgian consumers (%)
Aid	– Give clothes, food, medicines for developing countries	57.8	65.6
	– Give clothes, food, medicines for Belgium	56.6	71.6
Solidarity	– Financial support for the developing countries	48.1	24.7
	– Financial support for the poor in Belgium	29.7	19.9
	– Ethical investment	18.6	3.4
Justice	– Take part in a demonstration	30.1	5.4
	– Take part in awareness-raising campaigns	20.7	4.9

Source: Data are for 2004.

colonialism). It seems logical to consider that giving greater importance to external factors will increase the commitment to the South, while considering internal factors more important will tend to limit the commitment (on the *'what good will it do?'* basis). We also thought that identification of the causes would determine the types of commitments (aid, solidarity and justice).

Compared to Belgian consumers, Oxfam clients are more likely to cite causes that are external to the developing countries to explain why they are under-developed. Oxfam clients are more radical when considering the subject of dependence – debt of Southern countries, colonial experience or North–South relations; 43.3 per cent of Oxfam clients claim to be in total agreement with the view according to which unequal North–South trade is a cause of under-development of developing countries. Thirty-five per cent (compared to 17.9 per cent of Belgian consumers) completely agree that debt is the cause of under-development in the developing countries. Among internal causes, there is a general consensus that corruption of the political and economic elites in the countries of the South is a cause of underdevelopment. Apart from this item, between a quarter and one-third of Oxfam clients do not agree that internal factors are causes of underdevelopment. They seem to be more likely to give internal causes lower importance than external causes, as Table 9.5 shows.

Finally, we studied the type of commitment among Oxfam clients so as to define commitment profiles. To this end, we analysed replies to questions that correspond to our theoretical differentiation between the types of aid (solidarity,

Table 9.5 Identification of internal and external causes of underdevelopment (comparison Oxfam customers/Belgian consumers; several answers allowed, results in percentages)

	Do not agree		Agree	
Internal causes	*Oxfam*	*Belgian consumers*	*Oxfam*	*Belgian consumers*
Climate	31.3	23.5	47.8	58.3
Demography	23.1	13.9	53.7	72.2
Corruption	4.8	2.3	87.2	92.6
Mentality	30.2	13.5	53.1	71.3

External causes	Do not agree		Agree	
	Oxfam	*Belgian consumers*	*Oxfam*	*Belgian consumers*
Colonialism	14.0	20.8	70.0	51.2
Debt	11.6	18.1	79.2	61.5
Dependence/economic trade	5.0	10.4	84.9	72.2

Source: Data are for 2004.

charity, justice) and compared them with identification variables. Among the clients who favour an *aid/charity relationship*, retired people and households in the 34 to 44, 55 to 64 and 65 to 74 age groups are over-represented, as are all the incomes over 1,000 euros. Among clients more inclined to *solidarity*, there is an over-representation of those who give high importance to external causes of underdevelopment and who claim to be to the Left politically. More specifically, clients who are involved in giving financial aid to the developing countries are better educated, more religious and practising churchgoers while having incomes above the average (2,500 to 5,000 euros and >5,000 euros). They also give high importance to the external causes of underdevelopment. For clients who favour *justice*, the profile seems to be more masculine, better educated and more to the Left. They also give high importance to the external causes of underdevelopment.

Enlarging the clientele?

The survey among Oxfam clients underlines the over-representation, when compared to Belgian consumers, of two types of profile: the young, and people over 45. This could confirm our hypothesis of 'committed' consumption, given that these profiles correspond to the stages in life when commitment could be more frequent (cf. the biographical availability issue: Rotolo, 2000; Siméant, 2001; Stangherlin, 2004). Furthermore, we also observed that commitment profiles among this clientele are split into a more political fringe (solidarity and justice), which is more supported by younger clients, and a more moral fringe (aid and charity), which is supported by older clients.

Regarding the necessity to enlarge their clientele, our results show that Oxfam World Shops can choose between several options. First, they could build on what they already have by targeting external communication on these two profiles, with a different message for each group (young/solidarity or old/charity). This option is preferable if one considers that the 25 to 45 age group are definitely less likely to be mobilized as they are caught between professional and family commitments. Perhaps communication aimed at families could be further developed, for example, by participating in family events in which children (including the youngest ones) are involved (Saint Nicolas, Christmas, Halloween). The message could be more focused on the ease in purchasing Fair Trade products, with an increased development of mail order policies or linked to new technologies (the Internet), for example (see De Pelsmacker *et al.*, Chapter 8, this volume).

Our survey has allowed us to reposition the commitment to Fair Trade as part of a more general commitment to distant causes. This allows us to explain that the Fair Trade organizations' – and more specifically Oxfam's – actions (or lack of actions) to increase the battalions of fair consumers are part of a slower process of evolution of our society and of their relations with the populations of the South. We discovered, for example, that 92.6 per cent of Belgian consumers and 87.2 per cent of Oxfam clients thought that underdevelopment was due primarily to the incompetence of the local political and economic elite. We were deeply concerned about these figures. They seem to indicate a significant consensus within Belgian society (without it being a national peculiarity!) concerning the responsibility of the elites in the South. These figures can be an obstacle because such a view of the responsibilities for underdevelopment could discourage and lead to resignation as to the impact of actions by consumers in the North in favour of development in the South. Cleverly, Fair Trade organizations' communications to the general public do not mention the involvement or not of the public authorities in commercial projects in the South. Even more striking is the fact that Fair Trade organizations present Fair Trade as a system that tries to eliminate all unnecessary third parties between the small producer and the consumer (see Le Velly, Chapter 14, this volume). Nevertheless, we think the problem is larger and exceeds the unique context of Fair Trade. The results of this survey rather suggest that it would be better to strengthen development education policies, which appear to us to be the necessary factor if public opinions are to evolve when regarding the current approach to North-South relations. We think that in the long term, a well-designed development education policy could be the vector for strengthening the commitment to the problems of the South and therefore, in association with Fair Trade organizations' communication campaigns (we are thinking mainly of Oxfam), for increasing the number of committed Fair Trade consumers.

Limits of the consumer citizen

Our study should reflect generally on consumption as a political behaviour. As Chessel and Cochoy note:

> *Taking the so-called* political consumption into account should be the first step towards the acknowledgement that consumption is, by its ways, a practice of political nature. . . . The market has always been an eminently political space: all choices, determined either by axiological or material purposes, are involved in the evolution of all relations of power and the definition of a common world.
>
> (Chessel and Cochoy, 2004: 9; trans. author)

If all consumption is 'political', there is a danger that should be avoided: the overemphasis of the role of these 'ethical' or 'responsible' consumers, especially in this context of the global economy. According to the German sociologist Ulrich Beck, this political consumer power is an essential responsibility of a 'global civil society' capable of establishing itself as a true counter-power of a global hegemonic capital. He pointed out that 'no strategy exists that allows Capital to answer back to the growing counter-power of the consumer. Even powerful worldwide groups cannot fire the consumer' (Beck, 2003: 34).

Serge Latouche, whose opinions are generally far from Beck's, also observes, undoubtedly with some bitterness, that the consumer:

is no longer willing to remain a passive user, a role that the consumerist system reduced him to, leaving to the syndicate and to the state (or what is left of it . . .) the responsibility of the counter-power in relation to the market. He is now asked to rediscover a form of citizenship in the very heart of the market dispossession.

(Latouche, 2003: 257; trans. author)

Buying organic, ethical or equitable products, rather than from large agro-food groups said to be less concerned about social, ecological and economic production conditions, is for these authors therefore a profoundly political act. This absolute weapon against the hegemony of the global capital is nevertheless subject to certain conditions among which is having a choice of alternative products and sufficient buying power to make this choice. In the Western world, these two basic conditions can be quickly fulfilled, but surely not by all. Some studies aiming to profile alternative consumers, such as customers of the Oxfam World Shops, show that these buying habits are still reserved to a certain socio-economic fringe of the population with sufficient social, cultural and economic capital. Our study seems to illustrate this association between a particular perception of socio-economic and political problems of countries in the Southern hemisphere (a situation that seems to be complex in the eyes of Oxfam customers) and the act of international solidarity through the purchase of equitable products. This study will most likely need to be prolonged by analysing, through a biographical approach, the accumulation of this economic, social and cultural capital that makes the engagement towards Fair Trade easier.

The second limit is mainly due to confusion between the consumer and the citizen fed by the supposed advent of a 'consumer actor'. Even supposing that the forms of alternative and sustainable buying habits do not represent an additional cost for households, it is necessary to overcome the usual obstacles of consumer mobilization. And these obstacles constitute the limits of a movement that would settle for calling upon the political power of the 'almighty consumer'. In contrast with the romantic reading of Beck, Marc Jacquemain (2002) clearly showed that the faces of the citizen and the consumer cannot be confused. The consumer mainly exercises an individual freedom: he or she alone decides to buy or not to buy a product free of genetically modified organisms (GMOs), a pair of shoes made by children in Thailand, or bananas grown 'in dignity' by a small Peruvian producer. His or her real power is limited to sanctioning, his or her arm is boycotting. It is the accumulation of individual decisions of consumers that can act as a counter-power of the world capital. Although consumer power cannot be denied in a system of democracy of opinion where the results of large commercial groups depend on their image, it can seem more uncertain to establish a true democracy on a planetary level based on such fragile movements. Citizen action, however, rests in the exercise of collective freedom, an active reaction in the heart of public space. By its collective aspect, it is also more costly and less directly accessible. It is *together* that we decide on the orientation of actions we undertake, that we reflect upon the consequences of one or the other act or word on the community and its environment. In no case, except if looking into the crystal ball of a few marketing wizards, can the role of the consumer extend beyond reacting to a positive or negative situation he or she is subjected to. The consumer is like a boxer, always waiting for an adversary to throw him or her a hard uppercut before reacting. The power of the consumer is far from representing a democratic overhang. At best it represents a limit, potentially protective against the drift of global capitalism. The issue is therefore not simply to know how to appropriately address the consumer in order to make money and coexist alongside large commercial groups. It is indeed necessary go beyond this aspect to besiege public space, to act instead of react. As Marc Jacquemain so accurately wrote, 'A society that builds itself upon the basis of consumers' reactions is less and less capable of analysing its collective aims' (Jacquemain, 2002: 69; trans. author).

We can conclude with Latouche that on the one hand, it is accepted that 'it is indeed by reacquiring the political power of the act of consuming that the citizen of an economically globalised society can hope to shift the course of things' (Latouche, 2003: 265; trans. author).

But in our eyes that certainly cannot suffice, and it would be naive or even dangerous to see this reacquiring of politics by the consumer as the only lever of citizen counter-power in this context of globalization. We feel that one of the main challenges of Fair Trade along with most initiatives of alternative consumption resides in the capacity to get out of the trap of the 'responsible consumer' that only represents a democratic option by default, rather than the outcome of a process aiming for the institution of a democratic system on a world scale.

Notes

1 This is the case for Oxfam, an outreach organization very involved in the movement for more equitable (Oxfam World Shops) and more ethical trade (cf. the *Made in Dignity* campaign) and an active participant in worldwide and European social forums. It was however not until Cancun 2003 and the World Social Forum in Bombay in 2004 that the Alter-globalization movement welcomed the Fair Trade movement.

2 However, the idea of consumption as a political behaviour (as a collective activity deployed within a public space aiming to influence public policies) is not so young. Marie Emmanuelle Chessel and Franck Cochoy (2004: 3–4) remind us of the importance of consumers' movements (from cooperatives to consumers' leagues) in Western countries at the end of the nineteenth and the beginning of the twentieth century (in parallel with the development of the first theories of the sociology of consumption).

3 The paragon of such an enticing text may undoubtedly be found in one of the last publications of the German sociologist Ulrich Beck (2003). This will be discussed below.

4 According to figures from Laure De Cenival based on an EFTA study, 'Fair Trade products are available in 70,000 sales points: 3,000 specialist shops but also 33,000

normal shops and 50 supermarket chains. Nutritional products account for 66 per cent of sales and coffee is responsible for half of this' (De Cenival and SOLAGRAL, 1998: 21).

5 Active awareness: the respondent provides an answer without the interviewer having to provide any prompts or choices.

6 IPSOS, October 2000 and October 2002; IFOP, July 2001 and January 2002; Alter Eco, May 2002 (Lecomte, 2003).

7 In a broad sense, that is, referring to themes that are more remote such as international solidarity, ethical trade and global craftwork sales.

8 Significant difference estimated by Chi^2 test with Chi^2 = 0.016. Among Oxfam clients, the 18 to 24 age group represent 4.7 per cent, the 25 to 55 72.95 per cent and the 55 and over: 22.35 per cent. Among the 'Belgian consumers', the 18 to 24 age group represent 2.68 per cent, the 25–55 80.0 per cent and the 55 and over 17.32 per cent.

9 Significant difference estimated by Chi^2 test with Chi^2 = 0.000.

10 Relatively significant in the Chi^2 test with Chi^2 = 0.035.

11 For example, employees with non-administrative roles represent 39.01 per cent of the Oxfam sample and 32.3 per cent of the Belgian consumers sample.

12 Qualified workers represent 8.9 per cent of the Belgian consumers sample and 4.05 per cent of the Oxfam sample.

13 Significant difference estimated by Chi^2 test with Chi^2 = 0.000.

14 This figure is too weak and could be suspected to be biased (only people responsible for the daily purchase of consumption goods could answer the questionnaire).

15 Significant difference estimated by Chi^2 test with Chi^2 = 0.001. e.g. people with low incomes represent 12.5 per cent of the Oxfam sample and 4.75 per cent of the Belgian consumers sample.

16 To be a little more exhaustive, we also observed that the Oxfam clientele is more Flemish- than French-speaking (Chi^2 = 0.021) and that they claim to be politically more to the Left (Chi^2 = 0.000) than ordinary Belgian consumers; 57.4 per cent of the Oxfam clients claim to be to the Left compared to 31.2 per cent of Belgian consumers.

17 On the other hand, there is no significant difference between the results obtained within our two samples about the 'ecologic' or even 'biologic' dimensions of Fair Trade products.

18 Thus, the ecologic concern seems to be less important among the 'committed' Fair Trade consumers than among the average Belgian consumers. This result could show a relative dissociation between the environmental 'cause' and the Fair Trade movement. Of course, further work should be done to confirm this supposition. Nevertheless, our study shows that Oxfam clients are more committed to environmental association than Belgian consumers (see below).

19 These results should be used with caution. The question was not precise (what is an 'active' member?) and may have influenced the results (cf. low level of no reply and no associative commitment).

20 These results should be looked at in conjunction with the comments in the market analysis concerning the probability of consumers changing.

10 What justifications for a sustainable consumption?[1]

Coline Ruwet

Introduction

At the beginning of the twenty-first century, there is growing worldwide awareness of the social and environmental impacts of daily consumption behaviours in the richest countries on the planet. The analysts who looked into this phenomenon generally agree that it is not just a fashionable issue but rather a 'structural trend of the change in consumption behaviour',[2] and a 'deep and lasting movement' (Fraselle, 2003: 76). Surveys conducted by marketing agencies or by centres specialising in consumer research are taking an interest in this new buyer niche and are attempting to determine their profile and their motivations.

In this chapter, I propose to answer the question of the 'whys' of sustainable consumption by focusing on the arguments summoned up by persons who have an increased awareness of this issue to justify their concern for the social and environmental impacts of their daily buying actions. The discourses of members of a non-profit-making association specialising in the networking of 'attentive' consumers and traders who give priority to human development, including social and environmental development and North–South mutual aid, were used as empirical material. Like Michelle Dobré (Dobré, 2002), I will place myself in the context of everyday life. This chapter will therefore deal, above all, with the consumption of frequently purchased products.

Unlike other contributions to this book, I will neither ask about the meaning of sustainable consumption, nor about the capacities – and incapacities – for action of individual consumers or consumer groups. I will apprehend sustainable consumption on the basis of the accepted meaning that is given to it by people who say that they are aware of this issue. Furthermore, I will not ask them the question as to how they put their convictions into practice. I have decided to use the word 'responsible' as a qualifier to refer to such consumers.[3] The definition given to the concept of responsibility in this contribution is akin to the one proposed by Hans Jonas (Jonas, 1979 [1990]): it involves a new form of ethics, an 'intentional duty'. 'Responsible' consumers are thus characterised by the awareness of their power to act and the need to regulate it politically and/or to give it a social meaning.

This chapter will be structured in two sections. The first will consist of proposing a typology of the different profiles of consumers aware of the issue of sustainable consumption and their logics of action. The second will set out the joint arguments – brought together in three priority areas – put forward by all those questioned to justify their interest in the social and environmental impacts of their everyday consumption.

Close-up of the responsible consumer: typology of buyer profiles and their logics of action

The consumers' profiles described here were constructed from an in-depth analysis of comprehensive interviews of members of a non-profit-making association specialising in the promotion of sustainable consumption conducted between January and April 2004. Founded in 1999 on the initiative of consumers, '*Ainsi fonds font fond*' initiated an overall reflection on the impacts of everyday consumption. Its aim is to provide concrete encouragement to business projects located in the Liège region in Belgium in which human development is a central feature by setting up a network of 'attentive' consumers. All those questioned, between 25 and 70 years of age and from heterogeneous economic and social backgrounds, were members and sometimes active members of that association – either sitting on the Administrative Board, or 'reporters' for SENS, the periodical of the association.

For twenty years or so, many studies, both in Europe and in the United States, have been devoted to the definition of the characteristics of the different profiles of consumers aware of sustainable consumption, whether those characteristics are cultural, sociodemographic, geographical or psychological (see also Chapter 8 and Diamantopoulos *et al.*, 2003).[4] Furthermore, research is being carried out to assess the potentialities for adopting consumption behaviours compatible with sustainable development ([Shove, 2005] for Belgium see in particular Bruyer *et al.*, 2004). In this contribution, it is neither a question of determining the extent of the 'niche' of buyers concerned by the social and environmental impacts of their consumption nor of describing their characteristics. In order to do that, a quantitative survey based on a representative sample would have been necessary.[5] The objective of our case study was to examine under a microscope the arguments to justify their behaviour marshalled by those who affirm that they are concerned by sustainable consumption. Even though the 'niche' of persons with awareness of sustain-able consumption often tends to be considered as a group with similar concerns, the interest of focusing only on those who have taken the step of becoming members of an association centred around that issue is to emphasise both the heterogeneity of the arguments they put forward and the points in common that bring them together.

The analysis of the representations of those questioned within the scope of my empirical research made it possible for me to establish that an opposition between a mode of production and 'industrial' products distributed in supermarkets, and a mode of production and 'craft' products connected with short circuits and local trade, was one of the central disjunctions which structured the discourses. On the one hand, the 'industrial' mode of production is intrinsically linked to 'traditional' – mass – consumption whereas, on the other hand, the 'craft' mode of production is associated with sustainable consumption. The comparison of the different ways in which individuals conceive of that opposition – expressed in the form of double-entry tables – enabled me to construct a typology. Two logics of representation correspond to the groups identified: those who envisage sustainable consumption as a means of protest and those who consider that it represents a credible alternative to the industrial mode of production. That typology is not perhaps exhaustive. Additional interviews conducted among other groups of 'responsible' consumers would probably make it possible to complement this typology.

Sustainable consumption as a means of protest

Despite the variety of arguments used to justify their interest for the social and environmental impacts of their daily purchases, what brings the individuals classified in this first group together is the fact of considering sustainable consumption as a means of protest. Consuming sustainable products is experienced as a way to question some of the principles at the basis of the market system but not its overall logic. The current economic system is not fundamentally questioned; it 'simply' involves adjusting the deviations, the perverse effects that it creates, by introducing the values associated with sustainable consumption. These consumers are trying to find a compromise, a balance between 'sustainable' and 'non-sustainable' buying.

The meaning given to sustainable consumption by the first group is expressed in their actions. These buyers are characterised by significant variations in their behaviours according to circumstances and situations. Generally speaking, the sustainable consumption approach does not induce them to make in-depth changes in their buying practices. They are 'happy' with certain changes in limited areas. These buyers consider sustainable consumption as an investment – in time, in money – to which they are ready to consent in order to adjust the drifts of the current economic system.

Three different buyer profiles match that representation of the sustainable consumption approach: 'the committed and supportive consumer', 'the tradi-modern consumer' and 'the consumer borne by the wave'.

The committed and supportive consumer

For the committed and supportive consumer, 'responsible' consuming is above all an act of citizenship. It is, on the one hand, a political act equivalent to a vote and, on the other hand, a charitable act with regard to countries in the South. 'Sustainable' products are basically perceived as political means of pressure or humanitarian aid intended to combat or mitigate the failings of the current system. The intrinsic criteria[6] of the products are of little importance.

This type of buyer considers that it is important to be concerned about the impacts of their consumption above all in the name of the collective interest – safeguarding the environment for future generations, helping poor countries and so on.

The sustainable consumption approach of the committed and supportive consumer is associated with the non-market sector – represented by non-governmental organisations (NGOs) – and is opposed to the market sector – multinationals. NGOs are characterised by their concern, or even their combat, for respect for the human being and his environment, whereas multinationals are only interested in profit and immediate profitability and, to achieve that objective, exploit the poorest people and create pollution. Consumers are created as a new counter-power to make up for the impotence of national governments and trade unions in the face of international economic actors such as multi-nationals. Solidarity is thus a central value of the discourse of the committed and supportive consumer. Furthermore, the approach of this consumer profile is only meaningful from the moment it is integrated into a collective process. In that way, belonging to a group which shares the same concerns and objectives is considered to be very important for this type of consumer (Figure 10.1).

Mode of production

• **NGO**	• **Multinationals**
• Respect for the environment	• Pollution
• Recycling	• Squandering, waste
• Respect for social conditions and help for poor countries	• Exploitation of the poor
• Primacy of the human being	• Priority to profit
• Sustainability	• Short-lived
• Freedom of choice	• Manipulation
• Effort	• Facility

Products

• Fair + organic products	• (Unfair + non-organic)
• Price +	• Price –
• Trust in people	• Trust in labels

Figure 10.1 Structural analysis: representations of the committed and supportive consumer

Note: In the summary tables + = element is upgraded; – = element is downgraded.

The 'tradimodern' consumer

The principal specificity of the tradimodern consumer is to dissociate clearly the mode of production in relation to the intrinsic characteristics of the products. On the one hand, the 'craft' mode of production – associated with sustainable consumption – is depreciated because of its marginal nature, its lack of 'rational economic behaviour' and its failure to respect rules of hygiene. On the contrary, the tradimodern consumer puts great trust in the effectiveness, performance and reliability of the industrial mode of production. It is termed 'Cartesian' and 'organised' and is associated with a set of tried and tested techniques that make it possible to produce in a much more rational way than in the past – substantial quantities can be produced at lower cost. On the other hand, overproduction associated with the exacerbation of the principles at the base of the market world may have negative repercussions as regards the intrinsic characteristics of products – quality, and respect for social and environmental conditions. The desire of this buyer profile is consequently to succeed in reconciling the positive impacts of sustainable consumption on the intrinsic characteristics of products with an industrial mode of production considered as 'economically rational' (Figure 10.2).

Mode of production

• **Craft (small-scale) –**	• **Industrial, Cartesian +**
• Marginal	• Normal
• (Non-professionalism)	• Professionalism
• No rational, economic behaviour	• Rational economic behaviour
• Deplorable hygiene/cold	• Clean/serious
• Ineffectiveness	• Effectiveness

Products

• Quality	• Non-quality
• Obligation to produce a particular result	• Obligation to use one's best endeavours (due care)
• Trust +, (traceability, labels)	• Mistrust (No traceability)
• Effort –	• Facility +
• Taste +	• Taste –

Figure 10.2 Structural analysis: representations of the tradimodern consumer

The consumer borne by the wave

One of the characteristics of the consumer borne by the wave is the vague and indefinite nature of his or her position in relation to the sustainable consumption approach. These consumers are made aware of the issue by one of their close relations and, at the outset, did not personally question whether or not their buying patterns had certain social and environmental impacts. Generally speaking, their system of representation is situated at the intersection between the first and second group (e.g. between those who consider that the logic inherent in the market world cannot be avoided and those who consider that a credible alternative exists). The consumer borne by the wave generally has a negative overall view of the principles at the base of the industrial mode of production. He or she sees it as a mode dominated by the demands of growth and profit – so that the consumer can 'have everything, all the time' – it sacrifices the environment and exploits workers – mainly women and children, i.e. the weakest. It is a world where 'making money' is uppermost and conviviality or a sense of service are jettisoned in consequence. A further element brings the system of representation of the consumer borne by the wave close to the system emerging from the second group, namely the introduction of a 'spiritualisation' of consumption. In that way, the disjunction mode of production, industrial products/mode of production, 'craft' product is associated with a death/life disjunction. A questioning in relation to health – as a result of the death of a close relative or an illness – is often the source of their interest in sustainable consumption. This approach thus appears – even in a still vague way – as a rejection of death and a progress towards life (Figure 10.3).

The system of representation of the consumer borne by the wave seems to induce him or her to deny the current logic of the industrial world to turn towards another logic. However, the constraints inherent in that mode of production – higher prices, greater investment in time, difficulty in finding the products – form a brake preventing him or her from wanting to get completely away from the traditional market system. No alternative therefore seems to be really credible to them as yet.

Sustainable consumption as an alternative: the inspired consumer

The second meaning given to the sustainable consumption approach is that of a credible alternative to the logic of the industrial mode of production, which is considered to be soulless, senseless and a negation of life. The inspired consumer considers that the way of life inherent in the sustainable consumption approach makes it possible to 'get away from the system' and to move towards another type of society, focusing on other values.

Whereas other surveys emphasised the presence of elements in favour of the industrial mode of production and considered that it was impossible to get away from the logic inherent in the market system, the opposition to be found in the inspired consumer between the mode of production – and the products – where 'craft' has a positive connotation and 'industrial' a negative one, is completely dichotomous. The craftsman – characterised by smallness and local background – makes efforts, respects the environment – does not create waste, respects the season cycle – and the social conditions of employees, puts his or her heart into the work, and works for the love of the art. The industrial mode of production – termed at one and the same time as being 'heavy', 'big and 'remote' – is destructive, pollutes, wastes, does not respect the social conditions of workers and is likely to stray off track – for example, agrifood crises (Figure 10.4).

Mode of production

- Craft (small-scale) +
- Close
- Respect for working conditions
- Effort
- Human demands are uppermost
- Respect for the environment
- Respect for the seasons, the natural cycle
- Conviviality (warmth), likeable personnel, sense of service

- Serious
- Not enough info
- (Marginal)

- Industrial – ('masters of the world')
- Remote
- Exploitation of women and children
- Facility
- Economic demands are uppermost
- No respect for the environment
- Aim = have everything, all the time

- You get a black look if you haven't chosen anything in two minutes, nobody to give you any information
- (Not serious)
- Bombarded with advertising
- Everybody goes there

Products

- Organic, fair
- Natural
- Health +
- Control +
- Transparent
- Trust
- Taste +
- Takes time
- More expensive

- (Non-organic, unfair)
- (Non-natural)
- Sickness
- (Control –)
- Non-transparent
- Mistrust
- Taste –
- Speed
- Less expensive

- **Life**

- **Death**

Figure 10.3 Structural analysis: representations of the consumer borne on the wave

The 'short circuit' or local shop with which the sustainable consumption approach is associated is very important in the system of representations characteristic of this second group. Indeed, on the one hand the origin of the product is known – the feeling of control over food is higher and, on the other hand, people know to whom they are giving their money. The short circuit is therefore synonymous with trust. Labels do not really inspire trust in the inspired consumer because he or she thinks that it is always possible to hedge or to get around the standards. Generally speaking, the negative connotation associated with 'traditional industrial products' is related to the race for profit. The industrialist is ready for anything to earn the most possible money and increase his or her power. The multitude of middlemen, the lack of direct contact between the producer and the customer give rise to a dissolution of responsibilities that can lead to deviations. The dichotomy between the craft

Mode of production

• **Craft +**	• **Industrial –**
• Small	• Big
• (Not powerful people)	• Power
• Small quantity	• Large quantity
• **De-growth +**	• **Growth –**
• Short circuit	• Remote
• Local shops	• Hypermarket, multinationals
• Makes efforts	• Facility
• Works while being proud of what he does	• No longer knows why he is working
	• Economics is the main thing
• Project for society	
• **Puts heart, love into work**	• **Will for power, race for profit**
• Qualitative	• Quantitative
• (Non-destructive)	• Destructive
• Responsible	• Dissolution of responsibilities
• Knows what quality is	• Plays on appearances
• **Environmentally friendly**	• **Pollution, wastage**
• Recovery, recycling	• Waste, overconsumption
• Hygiene –	• Hygiene +

Products

• Fair + organic products	• (Unfair + non-organic products)
• Natural, fresh, respect for the season cycle	• Artificial, chemical products
• Good, quality product	• Bad products, 'rubbish', stodge
• **Health +**	• **Contamination, disease**
• Price +	• Price –
• Taste +	• No taste
• Pleasure with an energy quality	• Compulsive pleasure, addiction

Associated feelings/values

• Respect, consideration for human being	• Suffering, exploitation, manipulation
• **Freedom of choice, autonomy**	• **Fear, dependence**
• **Solidarity**	• **(Anomie)**
• Awareness, thoughtful buying	• Unawareness, impulse buying/ compensatory
• Subjectivation	• Objectivation
• **Life**	• **Death**
• **Soul, spirit**	• **Soulless, spiritless**
• **Meaning**	• **No meaning**
• **Sustainability**	• **Short-lived**
• Use	• Acquire
• Trust +	• Trust –

Figure 10.4 Structural analysis: representations of the inspired consumer

and industrial mode of production has repercussions at product level. Products from the short circuit are synonymous with the terms 'quality', 'natural', 'fresh', while 'industrial' products are termed 'rubbish', 'stodge', 'artificial', 'chemical'. One of the main arguments in favour of products from organic agriculture is that they are 'good for one's health', whereas products from supermarket distribution are associated with 'contamination' and 'disease'. Furthermore, inspired consumers emphasise that there are two types of pleasure. The pleasure associated with eating organic or natural products consists of taking one's time to enjoy. It is a state, a feeling. The pleasure connected with 'traditional' consumption is compulsive – at food level, it is connected with filling oneself and, at material level, with having and possessing.

Next, the principal specificity of this second main logic of representation is to propose a coherent alternative to the principles at the base of the industrial and market worlds. Each of the two modes of production corresponds to a society. Contemporary society, product of the industrial mode of production, is a society that cultivates fear, insecurity and dissatisfaction. These factors influence (over)consumption, which becomes a compulsive and compensatory act. Responsible consuming means moving towards another type of society focused on other values. In that way, the sustainable consumption approach is assimilated with a lifestyle.

Finally, the greatest specificity of the inspired consumer is to spiritualise the approach of the responsible consumer. An intrinsically 'moral' feature is thus given to that approach. Death, soullessness and meaninglessness, dependence, unawareness, manipulation, are associated with the industrial mode of production, whereas the world of sustainable consumption is synonymous with life, meaning, awareness, freedom of choice and autonomy. Belonging to one world or to another influences 'the being' of the consumer. The 'ordinary' consumer is dependent, compulsive, unaware. The responsible consumer is autonomous, free, aware. In their discourses, inspired consumers explain that they went through a phase termed 'obsessional', 'extremist', 'fundamentalist', during which they were given over completely to responsible consuming. This attitude is often considered to be unliveable in the long term because it compels people to live on the fringe of society. Inspired consumers therefore consider that they can be autonomous and completely control the impacts of their purchasing behaviours by working on themselves. Sustainable consumption takes on an identity dimension in this consumer profile. Acting as a responsible consumer is perceived as the outcome of a new education in consumption. Great importance is attached to the overall consistency between their convictions and their actions. This is reflected in respect of a number of rules they have set themselves and by reflexive control over their everyday actions as a consumer.

Sustainable consumption as a means of protest or as an alternative: two representations of the relationship between the individual and the system

The reference to the life world and system concepts developed in the work of Jürgen Habermas[7] (Habermas, 1988) is particularly enlightening in interpreting the meaning given to the sustainable consumption approach among the persons questioned, and more particularly the central and recurrent opposition in representations between an industrial mode of production and a craft mode of production.

Generally speaking, the analysis of the representations of the individuals interviewed make it possible to state that the industrial mode of production corresponds to the system, whereas the craft mode of production is associated with the characteristic elements of the life world. From that point of view, the sustainable consumption approach may be interpreted as an attempt, in a context of crisis of meaning, to reintroduce values, a significance within an economic system that has completely dissociated itself from the life world. For consumers aware of the issue of sustainable consumption – and particularly for the inspired consumer – the consumer society would therefore no longer seem to fulfil its function of cultural defence against anomie (Jackson, 2006).

The two main logics of representation – the sustainable consumption approach as a means of protest and as an alternative – develop two different visions of the relationship between the individual and the system. The case of the inspired consumer seems to be a particularly interesting one to develop. In fact, faced with a feeling of the meaninglessness of consumer society, the inspired consumer is characterised by a will to exclude the systemic component for the benefit of a society which would be based solely on the life world. The sustainable consumption approach is thus associated with freedom of choice, autonomy, awareness. Inspired buyers consider that responsible consumers can take a reflexive and critical distance in relation to society and control the impacts of their everyday buying behaviours by working on themselves. Inspired consumers thus consider themselves as being autonomous in relation to a social 'alreadiness' and try through will alone and their awareness of having completed control over their actions, the repercussion of the impact of their buying behaviours on others, the environment or health. The 'obsessional' phases that this consumer profile goes through are symptomatic of that search for control. In such periods, those consumers are faced with the resistance of the social aspect of the system to their will – that is the reason why they reach the conclusion that it is not possible to be a totally 'responsible' consumer.

Wide angle on the responsible consumer: main lines of discourse

The arguments of the responsible consumer are constructed by questioning some of the postulates underpinning traditional economic theory on which the figure of the consumer is historically based. Despite the obvious differences in the content of the arguments put forward by the respondents to justify their approach, there are justifications common to all those questioned. These may be structured into three main axes: a social axis, an environmental axis and a political axis.

Social axis

Trust and respect crop up recurrently in the discourses of those questioned to justify their interest in the sustainable consumption approach.

In the 'classical' approach to consumption, the consumer and the producer are envisaged as agents whose interests are divergent or even contradictory but who finally find an equilibrium according to the law of supply and demand. This mechanism is characterised by the uncertainty it creates, both in the producer and in the consumer. In fact, since only interest maintains the social connection, the different actors are neither assured of the duration nor the quality of the relationship. How can the consumer be sure that the lure of profit will not induce the producer to take useless risks or to transgress certain values? Agrifood crises or the accusations of failure to respect human rights directed at some multinationals have created a climate of mistrust. Emphasis on respect for ethical values, concerns expressed for the quality of products, control embodied in labels, are arguments aimed at establishing a relationship of trust based on transparency and respect between producers and consumers. The values of trust and respect are therefore flourished as safeguards against the deviations brought about by the omnipresence of the logic of interest in the 'classical' approach to consumption. It means restoring the primacy of interpersonal relations in the economic sphere and emphasising the social dimension of trade. The issue of the social tie, of the relationship of oneself with others, expresses a need for 'reliance' (Rochefort, 1997) as a core concern for the responsible consumer.

First of all, the thorough analysis of the justifications put forward by those questioned in our research shows that one of their common points, especially as a result of a crisis of confidence with regard to the abstract systems of modernity, is the desire to replace impersonal trust – also called 'system trust'[8] – dominant in market logic by personal trust. The crisis of confidence in systems shows itself in a search for 'quality' products among most of the consumers questioned.[9]

> The starting point of trust is a human relationship in spite of everything, which we might try to term transparent. In this case, we are talking about the trust involved in the relationship between a seller and a buyer. I think that if people have the impression that the person is transparent, that he is trying to give a maximum of information, that he is trying to help, to advise . . . I believe that it is all those things that go to make up trust.
>
> (tradimodern consumer)

Next, even though consumption is traditionally apprehended in an individualist way, the responsible approach stresses the relational dimension of that practice. The responsible consumer is someone who is aware that his or her buying patterns have an impact on others – not only human beings but also the environment. Faced with that fact, those questioned give a central place to respect, whether respect for the other's working conditions, for future generations, for the environment but also for oneself. The figure of respect valued by responsible consumers is characteristic of the system of differentialist interaction (Martuccelli, 2002); it is above all envisaged as a recognition of the other person, of his or her difference and of his or her right to integrity.

> Ah, what are the vital ingredients of the notion of respect, in my opinion? Well, for example, respecting nature means respecting a heritage to be bequeathed to future generations. Respect also means respecting myself and not eating just anything. It is a matter of integrity. And then respect for others, personally I am very sensitive to economic exploitation. I really have the impression that we can act as a lever in a different kind of network of people. . . . And respecting integrity and therefore not imposing, not. . . . Yes, respect for differences.
>
> (inspired consumer)

Environmental axis

One of the major arguments put forward by all the consumers questioned to challenge the current logic of the industrial mode of production is an ecological one. Among the consumers questioned, two visions of the relationship of man to nature exist side by side: an ecosystemic vision and a neofusional vision. In the ecosystemic vision of nature, humans and non-humans are considered as interdependent elements of a global system. Those elements are connected to each other and fulfil a function – the equilibrium of the system represents the harmonious state. That representation of nature does not give value to fauna and flora in itself; its objective is to safeguard the harmony of the system. Practices which might have a negative impact on the environment may therefore be balanced by favourable actions in other areas. In that way,

> the systemic approach offers an overall picture for an omniscient agent looking down from above. It lends itself to an evaluation focused on experts who would have the required capacity for constructing and manipulating that representation, for establishing measures necessary for a balance sheet, for emphasising perverse effects.
>
> (Lafaye and Thévenot, 1993: 521)

The will to find a balance, a compromise between responsible and non-responsible buying behaviours, expressed many times over by those who envisage sustainable consumption as a means of protest, induces us to conclude that they have an ecosystemic vision of nature on the whole.

> If everyone has a fairly balanced attitude and carries out a number of responsible actions, I have the impression that we are going to succeed in making a certain amount of progress and I think that we can see it. The world on the whole is a reactive world all the same. Whenever people had a problem throughout the existence of our dear planet, there was a reaction and people found a solution.
>
> (tradimodern consumer)

The neofusional vision,[10] partly inherited from the work of Arne Naess on deep ecology, is characterised by a radical questioning of the domination of man over nature. The community of reference is extended to non-humans – animal and plant species. In the most advanced version, man is considered as a species among others who may therefore not demand more rights than his congeners. Human practices (e.g. consumption behaviours) should therefore do their utmost to work towards preserving the untouchability of nature. The inspired dimension associated with the approach of those consumers leads us to affirm that whoever considers sustainable consumption as an overall alternative to current market logic is characterised by a neofusional vision. For such consumers, the relationship of human beings to nature is not envisaged from the viewpoint of the mode of balance but is perceived globally. Each of one's consumption actions is geared to choosing the products that use the least amount of natural resources.

> When I was 25 or 26, I already used to think about the use of the car, pollution, the number of kilos that people stuffed into dustbins, that used to annoy me. Taking care not to waste water, not to waste heating, I have been paying attention to all that for years.
>
> (inspired consumer)

Political axis

Each of the two conceptions of the sustainable consumption approach presented in the first section of this chapter is connected with one political project. Those who envisage sustainable consumption as a means of protest adhere to the political project of sustainable development, whereas those who consider sustainable consumption as an overall alternative to the industrial mode of production are closer to the vision of society proposed by the advocates of a 'de-growth' perspective.[11] The indefinite nature and the similarities of these two concepts make it possible for the members of the non-profit-making association who were the focus of our case study – where the heterogeneity of the systems of representations and motivations was described in the typology – to find a certain amount of cohesion and homogeneity.

In its definition, sustainable development integrates equal shares of the economic, social and environmental dimensions. One of the characteristics of that concept is not to question belief in the possibility of growth for humanity as a whole – that is moreover the main element differentiating that notion from the concept of de-growth.[12] Now, consumers who envisage sustainable consumption as a means of protest are distinguishable from those who envisage it as an alternative on account of their conviction that there is no way to bypass the logic of the industrial and market worlds – where it is worth recalling that the joint objective is to increase production and encourage growth. In other words, people who envisage sustainable consumption as a means of protest are persuaded that the material benefits, the well-being of a society, can only come about from the increase in production of goods and services in a country but, on the other hand, they oppose the social and environmental repercussions of that phenomenon. The concept of sustainable development therefore makes it possible for them to reconcile that twofold concern.

> An attitude of responsible consumption means having a consumption behaviour where you know that it will have a positive, long-term impact for the environment and for people. The human dimension and the environmental dimension. Sustainable encompasses all that.
>
> (committed and supportive consumer)

For the theoreticians of de-growth (Bernard *et al.*, 2003), the path that Western countries should take leads to a democratically managed healthy economy (e.g. an economic model which, at least, would not touch 'natural capital' and which would be the result of individual self-restrictions). A strict application of that theory leads each individual to take the decision to avoid any form of superfluous consumption as best as they can and to adopt the most moderate possible way of life. Furthermore, there is a specific demand in our societies for a quality of life that cannot be satisfied by an increase in production of material goods. Aware of the off-putting, negative connotation attached to the idea of de-growth, the advocates of that theory stress the aspects such as happiness or enjoyment that the individual may find by deliberately choosing a simple life focusing on the relational, convivial and spiritual side. Breaking away from the economic imaginary of 'more is better' is the sine qua non for bringing us closer to that situation. Several reasons lead me to affirm that inspired consumers subscribe to the political policy of the advocates of sustainable decrease. First of all, for the individuals classified in the latter group, being a responsible consumer means, above all, consuming the smallest possible amount – not wasting energy, avoiding surplus packing, not buying things you do not really need.

> It is a responsible and more conscious choice. It means opting for decrease, that's obvious.
>
> (inspired consumer)

Next, inspired consumers give priority to local consumption – especially with the aim of doing their utmost to limit useless, environmentally harmful transport operations.

> There is the problem of air transport. . . . For me, for the moment, I have the impression that it is better to buy a Belgian product at the right price than to buy an African Fair Trade product because I am not sure that everybody has something to gain from it, environmentally speaking.
>
> (inspired consumer)

Finally, like the advocates of de-growth, the individuals who envisage the sustainable consumption process as an overall alternative to the market system stress the happiness and pleasure the individual may find by intentionally choosing a simple life focusing on the relational, convivial and spiritual side.

> It means taking that step . . . from the quantitative, from the material to the qualitative. It means changing from consuming to enjoying. It is really difficult to get into that behaviour, it requires an all-out approach.
>
> (inspired consumer)

Conclusions

Throughout this chapter, our analysis has concerned the arguments put forward by consumers, members of non-profit-making associations to promote sustainable consumption, to justify their interest in that approach. Generally speaking, sustainable consumption is envisaged as the manifestation of a disagreement with some postulates inherent in the consumption model constructed on the basis of classical economic theory. There are, however, obvious differences in the content of the arguments used. In the first section, a typology of the different consumer profiles and their logics of action were carried out. In the second section, we have emphasised three main lines of common justifications among all those questioned (Figure 10.5).

The interest of the study of the current trend of 'responsible' consumers, in my opinion, goes beyond the understanding of a change in the supply and demand of products or a transformation of the buying practices of a proportion – be it limited – of the population mainly in Europe and in the United States. That phenomenon may be placed in a relationship with a series of issues, crises and challenges facing contemporary societies. The emergence of sustainable consumption may thus be considered as the outcome of the disappearance of certainties regarding the foundations of industrial society. It opposes the deviations and the perverse effects brought about by the logic of the industrial mode of production, and rejects the primacy of the instrumental rationality on which the 'knowledgeable' definition of the buyer is based. The characteristic feature of 'responsible' consumers is their awareness of their power to act and the need to regulate it politically and to give it a social meaning. One of the

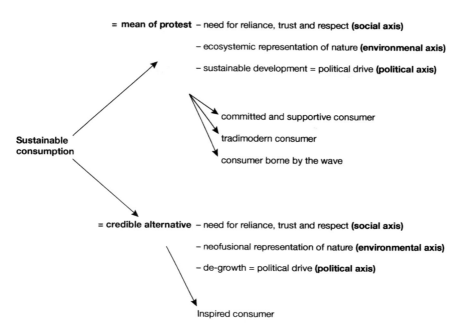

Figure 10.5 Synthetic diagram of the relationship between the different responsible
consumer profiles and the main argument axes of their discourses

specificities of the responsible consumer's ideology is to propose 'positive'
dynamics, unlike some reactions of fear and inward withdrawal. As a matter of
fact, it is not merely a question of contesting the worldwide deviations of real
capitalism: sustainable consumption promotes a project and proposes concrete
alternatives.

In a work entitled *L'imaginaire utopique aujourd'hui*,[13] Alain Pessin (Pessin,
2001) explains that, since the 1960s, we have been witnessing a transformation
of the utopian imaginary. Starting from the analysis of the contemporary
alternative imaginary, Pessin states that the content of utopia has changed.
Priority is now in the affirmative and no longer in refusal and combat. Building
an entirely new world is considered to be unrealistic and the current objective
is rather to propose pockets of resistance and make breaches. Whereas classical
utopia appeared as a coherent and comprehensive project, contemporary utopia
praises partial achievements. Contradiction[14] is considered to be inevitable.
In that way, it is no longer a question of people waiting for the Great Night
but rather of people acting on a day-to-day basis. Concrete experience rooted
in space and present time replaces an abstract and suspended time. Utopia is
apprehended as an ongoing construction in the *hic et nunc*. The project embodied
by sustainable consumption presents the characteristics of contemporary utopias
outlined by Pessin. Unlike revolutionary utopias, the action undertaken by the
responsible consumer does not claim to break with the industrial mode of

production and the capitalist system on a world scale from one day to the next in order to establish collectively a new system. It is rather a question of acting immediately, concretely, by carrying out 'positive' actions in the field of every-day life. The instituting capacity of individuals is emphasised. Is the sustainable consumption project sufficiently the carrier of collective aspirations and secured firmly enough in reality to mobilise a significant number of people and to have an effective impact on the rules of international trade on a world scale?

Notes

1 I would like to thank Edwin Zaccaï and Felice Dassetto for their comments on this text. I accept full responsibility for errors or omissions.
2 Interview of Robert Rochefort in 'La consommation citoyenne', *Alternatives Economiques*, Special issue, No. 10, March 2003: 10.
3 Committed, citizen, supportive, attentive consumer or Consum'Actor . . . all these epithets actually refer to a same buyer 'niche'.
4 This article included a review of the scientific literature of the research into this subject carried out over the past twenty-five years.
5 Every year, the CRIOC – Research and Information Centre for Consumer Organisations, Belgium – publishes a barometer of the behaviours of Belgian consumers in which the proportion of persons stating awareness of the social and environmental impact of their daily purchases is measured. See in particular Vandercammen (2002).
6 Price, freshness, quality, appearance, taste, smell.
7 In the theory of communicational acting developed by Habermas, society appears both as a system and a life world – there is a dialectical relationship between these two levels. Two types of integration and social reproduction coexist: the system attends to its material reproduction and is governed by an instrumental rationality, whereas the life world attends to symbolic reproduction and is governed by proce-dural rationality. Habermas considers that the pathologies of modernity can be explained by the colonisation of the life world by the system, i.e. by an extension of impersonal systemic mechanisms to regulate the social world. This phenomenon gives rise to a legitimacy crisis matched by a feeling of meaninglessness in individuals.
8 This form of trust studied by Anthony Giddens (Giddens, 1994) consists of delegating a part of one's capacity for action to systems and is particularly important in ensuring that they work properly.
9 For details of the importance and the meaning of trust in the economy see Thuderoz *et al.*, (1999).
10 For more information on sustainable development see Zaccaï (2002); on de-growth see Bernard *et al.* (2003), Latouche (2004) and Latouche in Chapter 12 of this volume.
11 This vision of the relationship between man and nature is termed 'neofusional' because it is a return to a renewal of a fusional attitude towards nature – man is immersed in nature, he does not act on it – characteristic of 'traditional' societies.
12 Sustainable development is an oxymoron for the advocates of decrease, since sustainability is only possible from the time when growth is blocked.
13 *The Utopian Imaginary Today*.
14 Emphasised by Edwin Zaccaï in the Introduction.

Part III

Dynamics of sustainable consumption

11 Consumption

A field for resistance and moral containment

Michelle Dobré

Introduction: how consumption leads to resistance

In the field of critical sociology, the topic of consumption is the arena of many determinations, the ground on which the consumer's 'freedom of choice' is easily assimilated to a veil of illusion, if not, as in the critical speeches of the 1960s, to a definitive renunciation of individual self-determination. Consumption as a field of determination and constraint may seem contradictory to the idea of searching for social capacities of action that would express, through the refusal in the action of consumption ('frugality', lesser consumption, self-control and self-limitation), the possibilities of 'ordinary resistance' that would be as many sources of creativity from daily action (Jonas, 1999).Yet, following De Certeau, we must attempt to learn more about those abstract 'capacities' that are more often presupposed than demonstrated. If many studies have shown the tactics of resistance through the diversion of production aims in the fields of work and popular culture (Coriat, 1988; Scott, 1997), the areas of consumption and daily life remain to be explored.

I will discuss each aspect of this examination in order to clarify the notion of 'ordinary resistance', a conceptualization I propose in order to analyze one of the ways to act in daily life. It is a form of practice among many others that reinforces itself and becomes subject to analysis only in precise historical circumstances. I advance the hypothesis that ordinary resistance is a modality of the practice that develops when the balance of power between the relational sphere (the 'social', where the relations with others unfold in all their forms, from the intimate sphere to the reception of the stranger) and the institutional sphere (the reticular 'system' of organizations and technical prescriptions) shifts durably in favor of the expanding institutional system. The relational sphere is thus reduced to the free reproduction of itself, that being in the general framework prescribed by the institutional network which tends increasingly often to predefine interactions while also being, *ipso facto*, that which makes them materially possible. The example of the confinement of children in the vast middle class of Western societies well illustrates this logic. Nowadays, practically no spontaneous interaction is possible between children who don't know each other, simply because the remaining occasions, non-mediatized by one form of institution or the other (e.g. school, parents' relations, associations) are no longer offered this type of interaction, still quite common a few decades ago. Thanks to individual automobiles, televisions, computers, electronic games and the institutional supervision of activities, one can reside in a neighborhood where many children live and are never seen by anyone outside their family circles. In its connection with the institutional sphere, the relational sphere finds itself dominated. It is when negotiation and/or open confrontation, classic modalities of action in public spaces, are no longer possible or efficient for the deployment of inherently social relations, that daily actions take on the appearances of 'clandestinity': one approves the mandatory character of the logic of the whole, the general framework, but diverts from it the sense and modalities to the advantage of values inherent to the 'world of life' (Habermas, 1988). It is a way of giving an account of the 'domination' different from Bourdieu's, emphasizing not reproduction but the capacities of social and daily actions to elude the different logics of power, including that of economic power.

Let me first specify in what sense I use the word 'consumption'. It is to be understood as mercantile consumption; that is, as consumption of which monetary resources are the condition of realization. Therefore it is not, at first, a 'social action' but an essentially economic one, of which the social dimension is either marginal or leaning towards marginalization. Thus, for example, we speak of 'mercantile consumption' in the case of a soup kitchen insofar as traders or citizens have given money (or goods that are always equivalent to money) so that those who are not solvent can feed themselves properly. Even in this case, it is easy to perceive to what extent anthropological resources (here solidarity, community feeling, charity) mix with the mercantile aspect of the final act of consumption. In my argument, I will separate consumption in the mercantile sense from consumption in its anthropological sense (the use with symbolic, social ends) of material goods (Douglas and Isherwood, 1979). In any case, the symbolic and anthropological dimensions are never entirely absent from commercial exchanges (but they can be reduced to a minimum, e.g. greetings and expressions of thanks – which machines can very well reproduce). The reason I display interest for consumption mainly in its mercantile dimension is because it appears to me that its influence on our ways of living and thinking has reached an importance never seen before in other societies.

For each of us, daily economy really is mercantile consumption. Let me be precise that, through this argument, I was aiming to theorize 'civil resistance', that is to say a form of daily political action that would no longer take place in the work environment, whose capacities to group contemporary issues together appeared enduringly in crisis. In a previous study (Dobré, 2002), I found that the refinement of introductory hypotheses has led to a reformulation of this idea. Civil resistance would be the only visible part of a vast reservoir of hidden daily actions that constitute as many resources for a whole panoply of actions that are not necessarily expressed in the public space. Ordinary resistance would be the practical equivalent of 'ordinary knowledge' (Schutz, 1987) for social thought: a reservoir of practices as invisible as they are widely shared and that allow for a social life that is not reduced to economic rationalization (including, and especially, in commercial exchanges). Moreover, through various acts of boycott (Danone, Total), 'buy-nothing days' and other 'political' or 'infra-political' initiatives (Scott, 1997) by more or less organized consumers, one can legitimately evoke civil resistance with regard to consumption; but that is another question.[1] Ordinary resistance would thus rather be the inherently social resource of what can – or cannot – translate (according to modalities that are not easy to define, due to a lack of action theory that would not be the rational actor's unsatisfactory one) into open opposition, political action in the public sphere or in civil resistance. However, the precise form into which ordinary resistance will be embodied in the public space depends on a historical state of the balance of power between dominant and dominated; it is thus answerable to a theory of power or domination that would replace those of Foucault, as it would start from a specific transformation of power relations in today's world (Bauman, 2000).

In the following, I will focus my argument on two main issues. First, I will ask how to ground theoretically the ordinary resistance in everyday life, by confronting it with theories of practice such as Bourdieu's (see 'Ordinary resistance and daily life'). In 'How to observe the practices of ordinary resistance?', evidence from twelve focus groups conducted in France aims to raise a few questions about the observation protocols (statistical data, surveys) and to indicate some new directions for further research.

Ordinary resistance and daily life

We have seen that henceforth, daily life is the scene we need to observe. In our daily lives, mercantile consumption is of central importance. The economic transformations that have occurred in industrialized countries since the end of the Second World War went in the direction of an industrialization of consumption on the example of the work rationalization process that unfolded through the nineteenth and the beginning of the twentieth century. I use the expression 'industrialization' not so much in the sense of the advent of mass consumption as a way to compare systematization and rationalization of daily techniques to similar processes that happened in the field of work and production during the previous period. It is also during the postwar period that the semantic shift from 'housewife' to 'consumer', a less feminine expression (so valorized), has occurred, thus confirming the new importance the consumer actor was to acquire in the economic universe.

The criticism of consumption signifies its definitive and invisible inclusion in our social world

As is often the case in the context of modernity, the surge of criticism signals the end of a process and its ingestion in the 'normal change' (it is, by the way, a classic mode of operation of knowledge in social sciences, to pay attention to endangered objects). Baudrillard's essay on the 'consumption society' (1970) has not meant the death of the social form he was trying to dismantle; quite the contrary. The analyses of this essay, which remains astonishingly pertinent, have proven true and have come to be, including, incidentally, in the intellectual journey of the author himself. His critique of the effects of consumption on individual liberty and on social ties has become commonplace – without having had the least inclination to leave this type of society. On the contrary, through the apparent 'virtualization' of commercial exchanges, never has the world of material objects, sold or bought, been so prevalent (and, we might add, urgent) in our environment.[2]

From the historical criticism of consumption . . .

The history of consumption criticism, even though it has not always meant mercantile consumption, is much longer. It starts in Antiquity with the Stoics, and goes on through the Christian vision of Stoic philosophers (and the adoption of their condemnation of material satisfaction regardless of spiritual accomplishment). That history of consumption criticism continues up until the eighteenth century with the constant opposition, in our civilization, between material goods (evil) and spiritual ones (good). In this sense, critics of consumption who have flourished in the post-Second World War period, with Baudrillard as well as Marcuse, Packard, Illitch, Gorz, de Jouvenel and all the institutional movement which opposed, from the 1950s, GDP growth to a measurement of well-being, these critics, then, stem from a long historical inspiration. What is less obvious is the parallel history of consumption's rehabilitation that is linked to another side of the culture of protest, Pietism (Campbell, 1990), which contributes to an opposite cultural movement which allows goodwill and tolerance in the face of 'human weaknesses'. This door, slightly open to satisfaction obtained through earthly goods, is reinforced, in the nineteenth century, by the opposition of romanticism to bourgeois accumulation (and the injunction for immediate pleasure in order to resist it), that will accompany the establishment of consumption and of pleasure through comfort. These will become the norm (almost universal) in the culture of the twentieth century. Of course, this process is not limited to the history of ideas: running parallel to the cultural rehabilitation of consumption as opponent to bourgeois accumulation (and Christian condemnation of instant pleasure), the second industrial revolution multiplies the proposed goods, urbanization gets well underway, the first superstores make their appearance and, in the new urban and solvent middle class, 'modern' comfort finds its defenders. Cooperative movements, such as Charles Gide's early twentieth-century Consumers League, try to make people realize this new importance of mass consumption, of which the first signs were already perceivable. It was only after the Second World War, however, that the surge of such consumption generated the most acerbic critics, at about the same time as the last cultural restraints of a religious or philosophical nature, those that had for centuries impeded the spread of material wealth and comfort, were weakening.

. . . to the triumph of mass consumption

Criticism of the society of consumption has not only become common ground, it has also accommodated the society of 'mass' consumption quite readily. Strengthened by the debate around it, it has never flourished in so much general indifference. It must be stressed again that the volume of consumption grows constantly in developed countries, even if those countries consider with dread that emerging countries aspire in turn to adopt the same model of abundance of which the ecological repercussions are increasingly recognized, if not greatly feared. Despite the impressions of dematerialization that are inspired by the ever-diminishing size of our miniaturized technical objects, material consumption is nowadays more massive than ever; this being the case through its own individualization. It is now technically possible to produce as job lot, that is, in limited editions, in a sufficiently flexible way so as to anticipate the latest trends in fashion (according to the 'zero stock' procedure that causes, as does home consumption, a maximum amount of trucks on the roads) and to give to each consumer the impression that the product he 'chooses' is intended expressly for him or her, all the while producing these individualization effects for millions, even billions of consumers.[3] Thus it is not the 'mass consumption society' that has gone west, but the way in which consumption is adapted for the masses so that with the support of technical means, it has become possible to live in a society of mass consumption where each individual, as a perfectly atomized consumer, thinks him or herself different from others and in a position to 'choose'. It is not hard to demonstrate that this consumer, subjectively endowed with 'freedom of choice', is structurally just as much in a position of constrained choice as he or she is similar to millions or billions of others who share this condition. The nature of this constraint, however, is in no way comparable to those constraints which were an integral part of the daily life of most people from previous societies. Let us simply say that individuals are unaware of the effects and massive nature of their choices, even though they are encouraged to worry about the risks to the environment and to human health that result from their own way of life. It is often at this point that consumers become aware of the binding nature of the institutional system that defines both the ways of life and the possibilities of change.

Is the atomized consumer thus deprived of his or her capacities for action?

There is an obvious gap between what individuals think of their own consumer condition and freedom of choice, and the generalization of the objective constraint (including the cultural, but also and more importantly the material constraints) in today's consumption society. Despite this difference, I will follow De Certeau's very plausible hypothesis, according to which consumers are not the helpless and complacent beings that the critics of the consumption society try to depict in their disenchanted way. It essentially consists in looking at the 'consumption's production', at the product of consumption in terms of uses. As consumers, and not as atomized individuals, we, each and none of us, display 'operating logics' often very ancient, some 'art of doing' and practical 'tricks' that allow us to preserve the impression that we have a minimum amount of control over our daily actions while being aware of the strong degree of constraint in the systemic universe in which we evolve. Furthermore, consumption practices happen to be a perfect setting for this daily exercise of control and self-limitation. Far from considering themselves as devoid of room for maneuver, consumers display many tricks drawn from practical wisdom and use them in order to counterbalance the integrating power of the mercantile consumption system. The style, as De Certeau says, is that of:

> moral resistance, that is an economy of the 'gift' (generosities for which one expects a return), an esthetics of 'tricks' (artist's operations) and an ethics of 'tenacity' (countless ways of refusing to accord the established order the status of a law, a meaning or a fatality).
>
> (De Certeau, 1980: 46)

In order to presuppose ordinary resistance in daily practices of self-limitation, we must accept the fact that the actor possesses, at least in part, some control over the sense of his actions. We must also invoke daily creativity that is no less than the permanent negotiation between the acquired and the new, imposed by habitus, for instance. If habitus equips the individual with a set of conducts inherited (for the reproduction) and to be invented in new situations (for the sudden appearance of what's new) that is in close connection with the group of origin, we would therefore have to ask the question of which habitus will be more favorable to the game with the daily constraints. One could bet that it will not be those of the privileged classes first, but rather that of the popular and average classes. However, if wealth dispenses privileges, that of reducing, up to the point of making it disappear, the daily aspects of life (the routine, repetitive and laborious aspects of living) are not the least. And that is what feeds the public dream of tabloids.

In spite of his lack of interest in daily life, the sociologist who most approached a conception of ordinary resistance in the emancipatory project of his sociology is Pierre Bourdieu. In a critical surge towards the defendants of popular culture (which replaced the tradition of research on the working culture in France), Bourdieu, however, explicitly mistrusts the temptation of researchers to see capacities of action where the structure imposes its powerful laws:

> Thus, the freedom of play that the agents have ensured (and that the theories known as 'of resistance' hurry to celebrate), in a preoccupation for

rehabilitation, like evidence of inventiveness, can be the condition of their contribution to their own exploitation.

(Bourdieu, 1997: 244)

In this context it is all the more surprising that Bourdieu, a sociologist at his best when it comes to revealing the truth about the various statuses, a practical theorist, saw himself invested with a mission of individual and intellectual resistance to the logic of domination, without recognizing in ordinary people even the least capacity, theoretical or practical, of going against this logic. Bourdieu has explained this many times in his *Pascalian Meditations* (1997) on this dichotomy, which he calls academic and which opposes reproduction to resistance without managing to rid himself of it (the title, in English, of his collection of essays *Contre-feux 1* is, after all, *Acts of Resistance*). On the contrary, by holding the noble role of resistance to the drastic conditions of the access to undeniable objectivity that a pulpit with the College of France begets, he uses a double tradition related to resistance, which precisely represents an obstacle to the theorization of ordinary resistance:

- *The resistant is an individual, a hero.* An old tradition, which dates back to Socrates and to Stoical clerks, takes part in the cultural process of individualization and heroism of resistance (with the image of Antigone, of Gandhi, but also, in its own way, of the Stoical philosopher, rich but as frugal as Sénèque, for example). In the literary and political fields, this tradition proposes to us figures of resistants (up until De Gaulle), exceptional individuals who incarnate to some extent a collective power, redeem it, represent it, and also, in a certain manner, replace it. Thus the handful of resistants in France has redeemed, symbolically, the collective ideological rout of fascism. Closer to our time, a figure such as José Bové, made a hero of his time, by the media, who makes it possible to redeem massive practices of industrial agriculture that are infamous destroyers of the environment and human health. More interesting still, one can suppose that the alter-globalist movement will be able to carry some weight in the international political sphere when it obtains (or will have been given by the media) a figure of resistant, individual and hero, as it should be.
- *The resistant is (inevitably) an intellectual* (this proposal is reversible). This more recent tradition depends on the first and has been reinforced in the nineteenth century, with the figure of the writer and the independent artist opposed to middle-class interests and consequently above party politics. In regard to Bourdieu, one will find his conception of his own role of intellectual resistant in the book on Flaubert (1992), where a thousand alarm bells are rung about the loss of independence of the intellectuals in the current context of subordination of artistic values to commercial values (which amounts to selling off the historical heritage of the empowerment of nineteenth-century intellectuals, represented in the figures of Flaubert, Baudelaire, Zola, and so on). Within the ranks of the Revolutionaries also,

the intellectual as an individual has found his place (the legend of Lenin getting into an armored train, for example: would the Russian Revolution have taken place if it had not been there?). In spite of the difficulties inherent to revolutionary theories, to start with Marx's, to grant a dynamic place to individuality, the figure of the revolutionary remains individual and Christ-oriented (Che Guevara, Sub-commander Marcos who, moreover, was originally a sociologist). It is because he does not enter the round of material interests, because he does not pursue power and material glories, that the intellectual is credited (and credits himself) with this critical and resistant capacity, that, in our day, ordinary people cannot afford. Bourdieu also subscribes to this tradition by speaking, almost on the mode of the idealization of the practical reason, of the one which, when taken into action, minding his own business, has neither the leisure nor the capacity to take a step back and analyze his own situation; far less extract himself from this situation permanently. Do we need to explain that this design of total absorption in the manual duty of the simple man is a pure intellectual phantasm? That very often, nobody better than those who are stuck with the material tasks, have better knowledge of the ins and outs, as many peasants commenting on the world market and perched on their tractor, have proven.

Resisting sweetness

All these thoughts put us on slippery ground in sociology: that of the 'consciousness', or worse, of the 'awakening', the sudden awareness. It has been mined for a long time, after the relevant criticism of the 'Subject'. Fine, if what one is likely to be opposed to is clearly perceived as harmful; one opposes it, justly one revolts, one calls upon the instruments of organized politics. But in the case of ordinary resistance in the field of consumption, another question arises: How to resist that which does us good? That which wants our own good? Because the improvement of material well-being is far from regarded as a scourge by the majority of the inhabitants of developed countries, even if we can clearly perceive what we have lost because of material progress (Boy, 1999). To speak of ordinary resistance in this context thus supposes that one runs culturally against these two relatively strong traditions.

The cultural device for heroism of individual resistance hides the nature of ordinary resistance, which is social

Works by historians about the Resistance during the Second World War, such as those of Semelin (1989), also went in this direction. Having explored in all their aspects the various resistant organizations and their actions, historians ended up being interested in the social conditions of the possibility of these actions. Here they were faced with the 'obscure' forgotten: women, village solidarities, the invisible cultural infrastructure that made possible the organized resistant action. In the same way, it is highly probable that the individual and heroic figure of the resistant has become hypostatic, metaphorically but also in a comradeship-like manner, a truly collective and social resource – without being able to be perceived as such by other individuals. There is a lack of knowledge here, but its mechanism is more complex than the work of the structure. It is historical individualism (and then ideological) that feeds this misconception of the social, and not individual, character of ordinary resistance as a resource of daily action.

The intellectual does not have the monopoly of resistance

Ordinary resistance, if it must take form in the 'relegated' space of daily life, is translated in the logic of practice by *reflexive acts*, consciously directed by the concern of withdrawing itself from mercantile logic (one could however stretch this reasoning to the influence of bureaucratic, technological or urban systems, and so on).

The practices of ordinary resistance are not, therefore, to be regarded as rational in the sense of a calculation. They are induced by requirements of two orders, both contradictory with the mercantile logic. The first requirement is inherently social (*sui generis*). It comes from the need for perpetuation, by itself, of the societal, and not through exogenous logics, whether we speak of technical organization (Lianos, 2002) or of the logic of commercial exchange. Maintaining an inherently social logic in human relations requires a distancing (which is also a diversion) from technical or mercantile logics, which tend to constantly replace the specifically social definition of human relations in the context of today's modernity. The second requirement is of a cultural nature. It comes from the autonomy project of the modern individual, a project that requires a minimum amount of control over one's own life or at least of one's biographical trajectory. Among the alternative projects of development that date back to the end of the 1960s, in line with the critique of the mass consumption society, the ecologists (in a very large sense) proposed a political project that could have allowed the 'atomized consumers' to become aware of their capacity for collective action in daily life, and to transform insulated acts, gone through individually, of mercantile logic refusal, into political change. However, this did not work – which raises another important question whose answer is not easy to find: What if ordinary resistance, which is a refusal of the acceptance of the general framework, which is trick, craft, diversion and a daily way to deal with things as they are, was not necessarily an engine of action in the political sphere, but rather the opposite?

I propose to shelve this question, knowing that the provisional answer is ambivalent, like the concept of ordinary resistance itself. In other words: ordinary resistance can constitute the breeding ground where a political project of much vaster change would be fed; but it can also conservatively allow things to continue without too many difficulties – precisely so long as these practices can give the impression of avoiding, as an individual, that to which all succumb collectively, namely with the logic of the great industrial and mercantile systems. It is also difficult to indicate the conditions in which the graft of a political project that would gather those individual acts (of collective nature) could hold. Would this be the advent of cumulative alimentary catastrophes, or something else? Would it be a clarification of the alternative political programs or the 'maturing' of opinion? I do not know. I, in any case, refuse to speak of an awakening, in the sense that consciousness is already there, and that the concept of ordinary resistance can perfectly continue without 'the subject'. The theory of ordinary resistance is not an agency theory, but a theory that seeks rather to seize the resources of daily life in the sense of Giddens (1984) ('gray zones' where the actor and the system meet). It is not necessary, in this perspective, to oppose practical reason to theoretical reason. The daily practice such as I understand it, in ordinary resistance, mobilizes knowledge just as well as action. Only, it is a question of interstitial knowledge and actions which do not take place on the ritualized scene of confrontations in public space. Again, to use an expression from De Certeau, ordinary resistance concerns those who, because they don't have their own place, must occupy (in thought and action) the place of others.

It is the societal, not the individual, that resists

We can speak of resistance because, in the case of consumption, that which we would like to avoid does not present itself to us straightforwardly, like a threat to be rejected, but on the contrary, like a good, an improvement and a progress. The extension of the institutional canvas (in which all the commercial acts narrowly overlap) takes place in each one of us in the name of 'more and better': more options, more effectiveness, more objects, greater comfort, better service and so on. The question of refusal that the concept of resistance implies thus cannot be asked in terms of the true and organized opposition, nor even in terms of civil disobedience (Thoreau in Bode, 1977), or of revolt, of the collective action in the traditional sense of the mobilizations. Or, to state it differently: comfort and material benefits of consumption appear to each other as necessary (even mandatory) and desirable, but not their ecological and human consequences.

It is the disproportion of the forces between daily action and systemic action and its representatives (the market) that invites us to explore ordinary resistance rather than the other forms of social action. There is not, and there cannot be, a confrontation of equal forces between the daily world and the economic, technological and technocratic unit that governs it (abstractedly indicated by the 'system'). The disproportion of the forces involved calls for tricks, for the *metis* (Détienne and Vernant, 1974), or other forms of action not yet identified. In this sense, ordinary resistance is indeed the fact of the dominated that recognize themselves as such. Women often practised this kind of slyness to bypass the obstacle of male domination. The division of forms and places of power, which Foucault calls 'biopower', Deleuze 'the holding company' and Lianos the 'new social control', must correspond to the new forms of confrontation in daily life. Foucault (1982) spoke about resistances, in plural, in one of his last writings. The 'new social control' of Lianos starts from the point that social forms of control, previously 'social', are precisely being replaced by technical devices that pre-configure our acts without us thinking of reacting (for example, if we observe our series of actions in a supermarket or on the subway, we realize that we cannot reconsider our steps in just any way, or come and go as we please, that the circulation and communication lanes in their broadest sense are pre-configured by the reticular technical device, which finishes that way by having a moral value – in the sense that morals dictate the ordinary codes of conduct).

The immediate consequence of these analyses for daily sociology is the report of a double epistemological and theoretical difficulty to apprehend the phenomena, to build the facts that best illustrate our approach. As soon as the concept of ordinary resistance is evoked, it awakes many associations drawn from the experiences of each one of us. I have noticed this quite often. The particular difficulty of the field and of the most conclusive facts is thus rather the result of the abundance of examples that one could give than of their scarcity. But systematic observation of facts of ordinary resistance in the field of consumption, whether frugality, alternative consumption or refusal to consume, is far from being able to be supported by institutional devices of data production, as is the case for consumption in the mercantile sense. In addition, the qualitative approaches are each time local and their value of generalization is lesser.

How to observe the practices of ordinary resistance?

Ecological consequences of consumption: the dead angle of statistics

For forty years, household mercantile consumption has been growing steadily in France, as in the other industrialized countries. Consumption grows just as much in quantity as in volume (in francs, then in euros). The structure of the stations of consumption changed inside the budget – without necessarily meaning a fall in the consumption of products that are polluting (containing oil, for example), exhausting non-renewable natural resources (still oil), or both at the same time. It is a challenge to try to determine the share of these products in public statistics, which do not have any specific entry on this subject. However, these are the simplest criteria of a durable consumption, which we retained in the investigation that I carried out for the first time in France in 1998 with the INSEE, on household ecological practices (Dobré, 2002). It is a similar investigation to the one carried out the same year in Belgium (see Bartiaux, Chapter 7, this volume). Only some practices, considered as revealing of a different attitude towards consumption, could be tested (for example, 'do it yourself', the deliberate use of the bicycle rather than the car, walking). On this subject, a similar report, which appeared in other investigations and on other subjects, is still verifiable: it is the educated middle classes that have the cultural and material resources to practise their ecological sensitivity. It remains that, if one can easily measure the overconsumption of such products (the individual car, for example, of which we can now measure the rate of multi-equipment, more than 40 per cent of households having two vehicles in France), one never measures – and with reason – what is not consumed. The refusal to consume (a-consumption), whether constrained or voluntary, does not appear anywhere in the description of the practices of consumption. But isn't it in fact a question of practices so minor, so tiny, that even the question of their measurement seems incongruous? To affirm that would be to subscribe, once again, to an idea of resistance reserved to some, the best informed (or most favored) who could afford it. In addition, it is not because a practice is not statistically observed that it does not exist. Indeed, it is because durable consumption is not measured (for lack of entry by the relevant criteria in public statistics) that one has the feeling that it does not exist. We must simply conceptualize consumption from this new point of view for it to come to exist in the 'reality' of the statistics.

In addition, in the more recent and more international context of the search for indicators of durable development, following Rio and Johannesburg, the problems of observation are even more acute. The fact of having connected, as ch. IV of Diary 21 had done, production and consumption in the same unit, resulted in focusing the attention on production, better statistically informed and easier to approach in the absence of sociological reasoning, to leave aside, once again, the too complex question of the ways of life. By reading the post-Johannesburg documents on the question of durable consumption (UNEP, 2004) and the list of operations to be carried out in the ten next years, one does not manage to carry great hope for the knowledge of consumption itself and the ways of life for a long time. The majority of the initiatives that are described in it are, still and always, detailed for and addressed to production.

Nevertheless, the current deficiency of descriptive statistical data of the consumption from the point of view of their durability moves us away from the possibility of establishing anything other than an approximate diagnosis on the ecological consequences of our mode of consumption. In any case, measurement has only a descriptive and diagnostic range. To see emerging, spontaneously, the question of voluntary restraint of consumption, the focus group's method proved to be more adaptable.

The experiment of uncertainty and the daily confrontation with consumption: the impossible search for frugality

Within the framework of European research that took place in four countries (France, Great Britain, Greece and Hungary) from 2000 to 2003 for the fifth action program of the Research Directorate of the CE, we approached the social conditions of the current experiment with uncertainty and insecurity. The concepts of uncertainty and insecurity were then reorganized in terms of 'social vulnerability' (Dobré and Lianos, 2002). Twelve discussion groups took place in each of the four countries, groups whose composition (screening) followed criteria related to the problems of socioeconomic uncertainty. The income/level of life criterion was homogeneous in each group. Certain groups were homogeneous from the point of view of age, sex, dwelling (remote or near the urban 'center' of the country, center/off-center). Others were homogeneous according to their predominant form of sociability in direct or indirect form. The anticipation of their trajectory in the future made it possible to distinguish between groups with a high level of aspiration (high expectations), and those with a low level of aspiration (low expectations). Each important criterion for the research was represented in two groups, to control for the specific importance of the factor considered in the course of interactions. All the groups covered all the significant social categories. The topics suggested for discussion were related first to uncertainty, in the vast meaning of socioeconomic changes and representations of the future. Next, the changes in several spheres, from the closest to the farthest, were evoked: work (precariousness), retirement plan, social protection, close relations (family, couple, friends and neighbors), risks, the financialization of the economy and European integration.

A question arrived at after the evocation of all these topics concerned the tactics that everyone could put in action in order to reconcile with an uncertain world, and this, with a general background of high uncertainty, who subjectively felt and in a distressed way by most of the groups.

Because the question referred to daily tactics, the topic of mercantile consumption slipped into the conversation without the organizer introducing it. Consumption, overconsumption, the constraint to consume, the permanent need to navigate in the world of commercial objects were topics that came up spontaneously in several of our groups in France. They appeared just as much in groups of which the members had an average or high income (craftsmen in Bordeaux; female executives in Paris), as in those with modest incomes (Lille, composed of men and women with modest incomes, working in declining textile industries) or very low incomes (unemployment, handicaps, off-center) in Indre. In all these groups a common point emerges: the world of mercantile consumption is lived as a difficulty, a daily constraint, and especially as being opposed to the world of human relations which, in parallel with the inducements to the consumption of objects, impoverishes itself and narrows. However, the modalities of the difficulty to consume we feel when invited to do so are different from one group to another. For the women, either in the group from Paris, or women from other groups, the difficulty is related to the education of children, to whom one must constantly show limits while at the same time allowing them to keep up with their peers. How, under these conditions, they wonder, can they manage to convey values to them? In the groups that come from popular classes, such as those of Nantes or Lille, the painful need to manage very modest budgets and thus to resist the urges to consume, including the fact of resisting indebtedness, credit, refusing to live beyond their means, combines with a nostalgia for bygone times when one could enjoy spare time without having to spend money, by riding a bicycle, gardening, or spending time with friends. In these groups, one tends to be opposed to the reduction of working time because it is said that spare time means time to spend, to consume, which one does not want.

At last, in the most deprived groups, like that from the rural district of Indre (center of France), exclusion results in the incapacity to take part in consumption, as well as by the criticism of this need in today's world. With the loss of employment, of incomes, of purchasing power, comes the loss of social relations: one cannot invite friends any more, and one is no longer invited. The isolation that results from this, far from causing proximity solidarity, makes these categories the most bitter critics of the economy, so much so as to make them wish for the disappearance of money. It is in this direction that tactics of ordinary resistance to mercantile consumption go: to privilege all that does not suppose the intermediation of money. To walk, to turn off the television, to gather herbs or fabricate toys for the children by oneself, to slow down the pace (to stop and reflect) – here is a whole panoply of measures that one proposes to withdraw from the influence of consumption, while remaining in the wishes register rather than in the one of effective practices. 'We should', 'we have to' are expressions that one also finds about the examination of needs. Instead of consuming, say people in several groups, it would be necessary to systematically ask ourselves what we really need.

The question of frugality, of 'how to keep ourselves away from the permanent invitation to consume', appears in the group of workers from Nantes in reference to the past, thus in a nostalgic and more passive way than in any other groups made up of individuals more engaged in the ascensional battle. In a group of teenagers who criticize consumption for the impoverishment of human relations that it induces, the tactics to reconcile an uncertain world refer to the spiritual quest, with the need for 'believing', whatever the form of this belief.

Conclusions: world without shelter and ways to resist

The aspiration to reduce our consumption, the idea that this is part of our means of daily action, the evocation of tactics to reach that goal, all of this is established in eleven groups whose composition includes the majority of representations on this topic in French society.

Only the group made up of individuals from the lower-middle class (Nantes 2) is characterized by a 'materialistic' hyperconformism, in comparison to the eleven others. It is exclusively in this group of small employees that solutions planned to face uncertainty and to reduce the feeling of insecurity depend mainly on professional success with, on the horizon, an increase in incomes, that is to say, of consumption capacities. The apotheosis of social success, as mentioned in this group, is managing to build a swimming pool in a second residence. This result, if it was to be confirmed in other researches, would be tremendous. The social layer (representing, what is more, the majority) carrying the ideology of consumption would be one for modest employees from the lower-middle class, which absorbed in France a good part of what was once called the popular classes – whereas, in all the other environments, consumption would be the object of criticism and distance, in speech as well as in practice. This result is corroborated by the statistics: we do indeed consume in enormous amounts while being unaware of the challenges and questions to this model, even if they are apparently present in many subcultures (the working-class culture, the popular class, the upper-middle class and the liberal professions are found, even though with different modalities, within this critique).

The capacity to imagine tactics of resistance, whether real or symbolic, is not universally widespread. I will especially stress the probable universality of the aspiration to do it, which is an established fact in light of the empirical data. The methods to do so, the degrees of realization as well as the practical engagement of ordinary resistance to consumption remain to be established on a case-by-case basis. It is because it is difficult to generalize, undoubtedly, that one could enrich the observation on this ground with thematic case studies.

This interrogation thus remains open, to know if the tactics of ordinary resistance can become the 'sand grain' of a material system that thrives through the deterioration of human relations (Haesler, 2005) – if, as I suppose in my theoretical premises, ordinary resistance draws from resources of direct sociality that are not individual, but properly social. That these resources come from a 'past' currently being brought down by the 'new institutional control' does not take away anything from the relevance of this questioning; on the contrary.

Notes

1 See, for example, Coline Ruwet, Chapter 10, this volume.
2 Fifty million objects are permanently offered for sale on EBay; a 'mercantile' way of ridding oneself of surplus goods without having to throw them away.
3 The European Environment Agency concludes that this is how the efforts of clean technologies and the gains in dematerialization to reduce the ecological repercussions of consumption have been, in fact, canceled out by the parallel growth in volume of European consumption in the past ten years.

12 Sustainable consumption in a 'de-growth' perspective

Serge Latouche

With globalisation, consumers (whether a dependent child, mother and homemaker, male or female worker, farmer, old-age pensioner) and end-of-chain buyers of goods made around the world must take on a new responsibility. They become *'consom'actors'* (active consumers), as some non-governmental organisations (NGOs) put it. It may appear strange, after denouncing economism and development (Latouche, 2003), to see us 'rehabilitate' the consumer, even consumption, even if they are *critical*. Consumerism effectively participates fully in the growth society that is responsible for the social, intergenerational and ecological injustice that spans the globe. There is no growth in production without unlimited growth in consumption sparked by all possible means, especially by the systematic manipulation of the consumer. The growth society may effectively be defined as a society that is dominated by a growth economy and lets itself be swallowed up by the economy. Growth for the sake of growth, through consumption, thus becomes the primordial, if not the only, goal in life. Now, such a society is not sustainable, because it exceeds the planet's carrying capacity and bumps up against the limits of the biosphere's finiteness, never mind the intolerable injustice that it engenders. So, after denouncing the farce of the market democracy, I propose to underline the challenge of responsible consumption, to explore the limits of Fair Trade, and finally to look at the paradox of deliberate simplicity.

The farce of the market democracy

Buying is voting. So be it. Yet this assertion does not suffice in itself to solve all problems. Ultraliberals and major capitalists have constantly repeated it, too, as they held up the 'plebiscite of consumers' to oppose decisions imposed on society by a regulatory state. The consumer is king, one might then say. The consumer's rationality is sovereign and the consumer's sovereignty is rational: on the one hand, consumers are the best judges of their own preferences and values. On the other hand, to take up a metaphor often used by economists, the sums that they agree to spend on such or such good or service look like so many stacks of ballots. It is not far from there to the statement 'Democracy is the market' (Vivien and Pivot, 1999: 51). The current 'pensée unique',[1] to use the

French expression, or single way of thinking, makes this leap. It has updated the demagogical slogan of 'small shareholders' that was thought up by the Popular Front's adversaries in France in 1936. It proclaims this hoax of a theme, along with the even more hoax-like theme of the small shareholders' referendum, loud and strong. It is the so-called 'shareholding democracy'.[2] 'An *ordinary voter* who has two hundred and fifty thousand ballots! That kind of democracy is the stuff of dreams!' René Passet points out in underlining the asymmetrical distribution of shares (2001: 30). According to such thinking, the homemaker would be voting in favour of the super- or hypermarket to the detriment of small neighbourhood stores, the disappearance of which would lead to the demise of urban life and a certain form of conviviality. She would also be the one to vote in favour of productivist agriculture to have *clean* or pasteurised foods at cheaper prices. She should thus bear the blame for the death of the countryside and disappearance of drinking-water, for polluting the water-table and soil with pesticides and chemical fertilisers. If we are to believe Monsanto, the Third World housewife is even clamouring for genetically modified organisms (GMOs) to escape famine![3] Finally, the users are allegedly the ones who ratify firings, downsizings, relocations, flexible wages and working hours in order to get means of transport that are ever faster, go ever farther and are ever cheaper; to have low-price clothing and electronics from emerging countries; to have cheap cars, mobile phones and computers, and so on.

According to this thinking, demand and demand alone is king. Its sovereignty is legitimate, because it is democratic and of the masses. All the fancy speeches about the benefits of markets' globalisation (felicitous globalisation) take up this refrain ad nauseam. 'The principle of the consumer's sovereignty . . . is . . . the principle of direct democracy practised in a referendum' is how the economist naturally puts it (Williger, 1999).

It is vital to deflate this arrogant argument, which is made on behalf of consumers but actually emanates exclusively from the major corporations' lobbies. In any event, these are not the voices of consumers as they are heard through consumers' associations.

The challenge of fair and responsible consumption

The citizen is also a consumer. Consequently, the consumer is a citizen. Political discourse considers the citizen to be sovereign, whereas economic discourse asserts that the consumer is king. The *consum'actor*, as some NGOs call the consumer citizen, means to claim this twofold recognition of his or her supremacy and exercise the legitimate prerogatives that derive therefrom without restriction. How then can one refuse to let the consumer know what he or she is buying, to know the provenance of what he or she is eating, as agrifood companies and the World Trade Organisation (WTO) claim to do? Traceability is really the minimum one must offer to those who one claims to be sovereign. Now, this traceability is far from total. Behind the 'first cold press' label for olive oil, for example, the French consumer does not yet have any guarantee of origin or production processes. The European authorities refused point-blank to give such a guarantee for chocolate. As for GMOs, transnational companies are waging a no-holds-barred policy of the 'fait accompli' with the US government's blessing. They are demanding that labelling be refused so as not to distort competition! According to Sarah Thorn, Director for International Trade of the Grocery Manufacturers of America (a food lobby), stating on a product's label that it contains genetically modified ingredients would be tantamount to dooming it to remain on the supermarket shelf.[4] Moreover, the US Agriculture Secretary, Ann M. Venemann, has even garnered the support of the GMO lobby in Brussels. In 2002, she told the European Commissioner in charge of consumer protection that it would be difficult to consider measures to label genetically modified products as not being trade discrimination measures.[5] What better way of saying that the consumer king has no business knowing what is inside the box and making his or her choice prevail. All they have to do is pay![6]

The consumer's metamorphosis from a normally passive (and infinitely patient) citizen into an active citizen and consumer demanding a minimum degree of respect is seen occasionally in crisis situations, when housewives and users refuse to have hormones in their beef, genetically modified organisms or footballs made by child slaves, and are ready to go so far as to boycott such goods. After the WTO's condemnation of Europe for refusing to import hormone-tainted beef and in response to the US retaliatory measures against various products, including Roquefort cheese, the President of CNJA (French National Centre of Young Farmers) declared, 'The only thing they know is money. You thus have to strike them where it hurts, in the cash drawer.' It is a pity that it took a crisis and the dismantling of a few McDonald's restaurants to rediscover these solid truths and proof of 'dietary degeneration'. It is also by reappropriating political power over the act of consumption that the citizen of an economically globalised society can hope to have some influence over the course of events.

This new situation contains a clear ethical challenge. Trade union activity and political activism call for sacrifices and an often huge moral dimension, but they can appear to be warranted by very tangible interests, such as wage rises, working conditions and various perks. In the case of civic-minded or 'citizenly' consumption, the relationship between ethics and interest is inverted. Of course, it is in everyone's well-understood interest to save the planet and ensure that it survives, to have healthy products, and even to have justice prevail to avoid chaos, but those interests come up against other, more immediate interests that would have to be sacrificed to the longer term goal. So, it is indecent to ask the welfare recipient or minimum wage earner to pay 30 per cent more for an organic product, even 50 or 100 per cent more for a Fair Trade commodity. How can one suggest that France's suburban residents[7] give up their favourite supermarkets for less inequitable forms of distribution, when the latter do not even exist close by and traditional farmers' markets are even scarcer? Of course, short producer–consumer chains can be organised, including in urban areas, in the form of fresh produce baskets (*'paniers-fraîcheurs'*) delivered to subscribers, associations to bolster family farming such as the AMAPs (Associations pour le Maintien de l'Agriculture Paysanne) in France, and workers' gardens.[8] A few hard-line environmentalists or fellow travellers use such schemes. It would clearly be all to the good if such initiatives spread. Under pressure from the people, the public authorities can strengthen such initiatives and give farm policies a less productivist bent. Germany has courageously blazed such a trail. However, can these formulae be mainstreamed? And yet again, where can one find the equitable car, fair kilowatt hour and moral cubic metre of water?

Fair Trade's limits

'You go to the supermarket and buy a packet of spaghetti and, without wanting to,' observes Francesco Gesualdi, author of a responsible consumption handbook (1999), 'you are financing the arms industry, because the multinational from which you are buying spaghetti also has weapons plants. Or else, you buy a tin of peeled tomatoes and contribute to the exploitation of African workers, since the multinational from which you are buying also owns pineapple plantations. In other words, each time you buy blindly, you can turn yourself into an accomplice of businesses that pollute, mistreat animals, or do other misdeeds.' Under such conditions, buying coffee bearing the 'Max Havelaar' or 'Transfair' label rather than an ordinary major brand is perhaps a citizenly action. Buying fairly, that is to say, in theory, according to a fair price, rather than letting oneself be sold an *unfair* product at the market price is a way to assert politics' mediation in trade. It is thus also an assertion of solidarity with far-away and unknown partners without denying their existence or being indifferent to their fates.

Unfortunately, it is not easy for the consumer to be a responsible citizen, for both subjective and objective reasons: subjectively, because the said consumers' tastes and desires are almost totally manipulated through advertising and the solicitations of mass distribution; objectively, because our consumer has almost no choice, even if he or she were determined to behave along more citizenly lines. Environmentally, politically and ethically correct buying is usually a true obstacle course and an act of heroism, without a true guarantee of success. Such a choice does not even exist for most products. Where do you find a Fair Trade car, an ethical refrigerator, a washing machine that supports the masses, and social software? You can already count your blessings if *traceability* is extensive enough to ensure that you are buying a three-piece suit that has been sewn elsewhere than in an upscale women's prison in South-East Asia. Are we truly voting for the slavery of Pakistani children when we buy a pair of shoes bearing a major multinational brand? Are we subscribing to the destruction of cultural identity when we offer our relatives a package tour?

Finally, consumption that is called critical, responsible, or ethical and Fair Trade can appear to be oxymorons on a par with sustainable development. The need to free ourselves from consumerism is as great as, if not greater than, that of putting an end to developmentism. Slogans such as 'consume ethically' and 'buy fair' are contradictory and perverse, for they conserve the core problem, i.e. the consumption imperative. And so, the NGOs that are working for Fair Trade are confronted with these contradictions. 'Never consume!' would not be a bad slogan in a way. The good consumer citizen is, if not a dead consumer citizen, at least a citizen who consumes little. Of course, this should not be taken to be a call for a radical purchasing strike, but rather one for voluntary self-limitation of one's reliance on the commercial circuits, which does not exclude a debauchery of shopping on festive occasions. It is above all a change in attitude towards the way we procure things from others to meet our needs in exchange for what we offer them.

The justification for the educational provocation of the call for consumer mobilisation lies in the fact that purchasing power is effectively one of the last powers available to the people to counterbalance, in some cases, or to oppose the powers of transnational finance. The citizen can no longer content him or herself with being a passive *user*, which is the role to which the consumerist system reduced ordinary citizens, leaving it to the trade unions and state (or what remains thereof) to counterbalance the market. Henceforward, people are enjoined to rediscover a form of citizenship at the very heart of commercial dispossession. Against the quasi-totalitarian sway of the market, which assumes the right to claim to be the people's spokesman, resistance and revolt become necessary. It is no longer a matter of negating or circumventing the class struggle, as propounded in social economy plans, but one of taking on this struggle from a different angle.

The paradox of deliberate simplicity

In *Tools for Conviviality*, Ivan Illich extols 'the sober joy of life'. 'Today's "human" condition, in which all forms of technology become so overwhelming that joy is to be found only in what I would call a "techno-fast"' (Illich, 1973: 43).[9] 'The necessary limitation of our consumption and production and the cessation of our exploitation of nature and labour by capital do not mean a "return" to a life of deprivation and toil. On the contrary, they give rise to the liberation of creativity, renewal of conviviality, and possibility of leading a worthy life, provided that we are capable of giving up material comfort', Camille Madelain observes (2003: 242).

The search for deliberate simplicity or, if you prefer, a restrained life, has little to do with a form of masochistic mortification. It is the choice to live differently, to live better, in fact, and more in harmony with one's convictions by replacing the race for material goods with the search for more satisfying things of value. The rare families that choose to live without television are not to be pitied. They prefer other satisfactions to the ones that the magic window can procure, namely a family or social life, reading, games, artistic activities, and free time for dreaming and simply enjoying life.

To the extent that it is possible, it is even recommended to return to *making things oneself*. In making our own little yoghurts ourselves, as Maurizio Pallante advocates, we eliminate plastic containers and cartons, preservatives, and transport (and thus savings when it comes to petrol, CO_2 and waste) and gain bacteria that are precious for health. And of course, we take a sizeable bite out of the GDP and taxes (VAT and fuel taxes), which has all sorts of recessive knock-on effects on our institutions and demand (less plastic, thus less oil, thus less tax revenue; positive health effects and thus less medication, less need for doctors; less road transport, and thus fewer accidents, less need for doctors; and so on). This same analysis may be done on giving up buying bottled water that comes from afar and returning to tap water from a treated local aquifer. There we have a virtuous spiral of negative growth (Pallante, 2005).

There obviously exists a whole series of other means that are, in turn, instruments and objectives and all reinforce each other reciprocally. We can think of reappropriating currency through the use of local currencies, melting currencies or non-convertible currencies (such as meal tickets or holiday vouchers).[10]

Obviously, this path is usually taken gradually and cannot be taken for granted, given the great contrary pressures exerted by society. This path requires that we master our fears: fear of the void, fear of running short, fear of the future, fear also of not fitting the prefabricated moulds, fear of standing out with regard to the reigning norms. 'The choice is one of living today,' François Brune observes, 'rather than sacrificing today's life to consumption or the accumulation of securities without value, to the construction of a career plan that is supposed to bring satisfaction tomorrow, or to payments into an individual retirement account that is supposed to counter the fear of not having enough' (2004: 175).

On the other hand, one thing is certain: the *external* imperative of austerity, experienced as a moral constraint ('I'm driving at 130 kph. That's not good. I'd better slow down to 100', for example), is both ineffectual and often counterproductive. The necessary aim of cutting down and the only way to escape consumerist addiction involve a change of mindset that makes the requisite behaviour 'natural'.

More advanced pondering of the ecological footprint effectively enables one to grasp the systematic nature of 'overconsumption' and the limits of deliberate simplicity. In 1961, France's ecological footprint corresponded to merely one planet, compared with three today.[11] Does that mean that French households ate three times less meat, drank three times less water and wine, and used three times less electricity or petrol at the time? Certainly not. However, the strawberry yoghurt that we ate at the time did not yet include the 8,000 km of various trips involved in its production and distribution! Neither did the suits that we wore, while sirloin steak devoured smaller amounts of chemical fertilisers, pesticides, imported soya and oil. However, the change in mental visions, even if not a deliberate decision, nevertheless results from the many changes in mentality that are in part prepared by (counter)propaganda and example. Mentalities have to 'trip a switch' for the system to change. Breaking out of the chicken-and-the-egg-type circle entails triggering a virtuous circle.

Countering manipulation and redoing the world

It is clear that we are not going to erase with a stroke of the pen the economic powers' manipulations. The latter is impossible to ignore and we must be careful not to underestimate it. Nevertheless, the aim is indeed to redo the world, and the means to achieve this is indeed to counter the manipulation and brainwashing to which we are subjected. It is time to start the decolonisation of our imaginations, that is to say, to *de-economicise* our minds (Latouche, 2004), to realise that our consumption desires and a world view dominated by the inevitability of economics result from the insidious manipulation of a system and are not based on any genuine needs. Things were different, could be different and should be different. We must set our sights steadfastly on this ambitious objective with the ideal of Fair Trade; that is to say, economies and markets that are mediated by social or political considerations.

When it comes to North–South relations, it is less a matter of giving more than of taking less. If we in the North truly want to include a farther reaching concern for justice than the sole and necessary reduction of our ecological footprint, perhaps we should acknowledge another 'debt', the 'repayment' of which is sometimes demanded by indigenous peoples, that of Restitution. Restoring lost honour (restoring a pillaged heritage is much more problematic) might be achieved by joining the South in a partnership for negative growth.

Notes

1 This expression was launched in France in 1995 by the influential leftist monthly *Le Monde diplomatique*, to which I often contribute. The following sentence may, in short, suggest its meaning: 'But should one accept this nice vignette of "pensée unique", this suave legitimation of a new dictatorship, that of financial markets, politics will amount – and it largely does – to little more than a pseudo-debate between parties of government shouting out the minuscule differences that separate them and silencing the significant convergences that unite them' (Serge Halimi, May 1997, 'When market journalism invades the world', *Le Monde Diplomatique*, http://www.mondediplo.com/ [English Edition, Editor's note]).

2 'Indeed, there's nothing easier than taking a democratic metaphor that is constantly cropping up in financial circles to its limit. Is not exposing the economic plans (those of companies and economic policy alike) to market opinion commonly presented as a form of suffrage? And aren't investors' decisions in favour of engagement or disengagement the expression of a type of vote? And corporate governance practices take this analogy up a notch by giving the financial *politeia* its *agora*: the general meeting of shareholders. . . . False democracy in a society that has no common concerns other than assets: It is not certain that the *polis* of pension funds is a *cité radieuse* (*i.e.* utopia)' (Frédéric Lordon [June 2000], 'Fonds de pension, piège à cons? Mirage de la démocratie actionnariale', *Raisons d'agir*: 106–109).

3 This is not yet the case, but thanks to propaganda backed up by corporate funds and with the complicity of the local elites, this seems to be the right path. Mohamed Yunus, the 'poor man's banker', was already spreading the propaganda about Monsanto's seeds in Bangladesh. Several African farm ministers carry this

insidiously penetrating campaign forward. By offering surpluses of genetically modified grain to Africa's starving states, the US is stepping up the pressure even more. In Iraq, this has already been done in the wake of the US's military intervention!

4 Quoted by *Politis*, Thursday, 10 October 2002.

5 'Such measures,' she continued, 'could cost US industry millions' Frédéric Prat (June 2002) 'Europe et OGM: Bruxelles, le passage en force', *Courrier de l'environnement de l'INRA*, 46: 76.

6 Still, since 2004, a European regulation does require labelling of products containing more than 0.9 per cent GMOs. Products derived from animals that ate GMOs are not concerned. On the limits of labels and product information's influence on consumers, see Chapter 6 by Rousseau and Bontinckx (this volume) (Editor's note).

7 Major French towns are typically ringed by areas of (high-rise or single-family) housing developments and shopping malls where people of more modest means reside (Translator's note).

8 The magazine *Silence* (Ecologie, Alternatives, Non-violence), 9 rue Dumenge, 69,004 Lyon, is an irreplaceable source of information and addresses, along with *Nature et Progrès*, 68 bd Gambetta, 30,700 Uzès.

9 Translated freely.

10 For more on this point see the final chapter in Serge Latouche (2003) *Justice sans limites*, Paris: Fayard.

11 This means that if every person in the world consumed at the 'French level', it would need three earths to be ecologically sustainable.

13 Social change for changing the consumer's behaviour

Application of the actionalist theory to the issue of consumption

Nadine Fraselle and Isabelle Scherer-Haynes

Towards more sustainable modes of consumption

A notion of needs, including those of the poorest, is central to the concept of sustainable development. Everybody should be able to enjoy the benefits of growth. Therefore, social aspects and solidarity are essential for sustainability. Furthermore, because future generations should be able to have access to natural resources, the environment should be protected and maintained.

Because both the social and environmental aspects of sustainability question our modes of production and consumption, new sustainable economic models – including solidarity approaches – are developed in contrast with 'conventional ones'. However, new realities (business ethics, corporate social responsibility and corporate governance) raise the issue of integration conditions, of the complexity of consumption and of the attempts that have been made by companies to integrate environmental and social issues into their business strategies.

On the one hand, the complexity of consumption dynamics has increased during the past years. Consumption analysis has to take into account new criteria that correspond to the latest social developments, including the following factors:

*   The supremacy of the demand side over the supply side and the birth of the consumer as a sovereign subject who is keen on exercising his or her influence over the market. New sources of information and communication, increased possibilities of choice and new assessment capacities have modified his or her consumption. All this influences both the choice of the goods to be consumed and the way they are to be used (or consumed).
*   New types of dialogue between consumers and companies within the framework of market co-regulation or self-regulation.
*   Increased business responsibility, thanks to the influence of demand: market ethics, business for sustainable development, risk management and self-auditing in production processes (e.g. norms, quality management, product and business certification, traceability, labelling).

- The importance of brands and their influence on behaviour, thanks to the creation of behaviour models and to their role in important socialisation spaces such as family circles, schools, the street, leisure and workplaces.
- New alliances between groups that have contradictory interests such as consumers, minorities, unions and new forms of power relationships within the frame of market globalisation.

Since the end of the 1980s, with growing criticisms of social work conditions in the textile industry, environmental and social issues have also been taken into account in enterprises' decisions. This was reinforced by the movement on sustainable development, a movement closely connected to globalisation. Globalisation has indeed increased the complexity of business activities, and has led the companies to assume new responsibilities on a global scale, especially in managing the international life cycle of their products. For global brands, reputation has become so important that minor changes can have significant effects on company value and profitability.

The idea of corporate social responsibility (CSR) is not new. The first perspective of this concept was launched by Bowen in 1953, who defined it as 'the obligation to operate lines of action which are desirable in terms of values of our society' (Bowen, 1953). This view of CSR as a social obligation has been dealt with in later marketing studies stating that there is a wide range of social obligations, including economic challenges. So the concept has re-emerged as a business issue. Today, CSR is definitively linked to financial performance and management. The search for new sources of economic growth and competitiveness is expressed by the 'stakeholder value' and the development of social performance management systems.

In this approach, consumers play an important role as stakeholders. Because they fear a loss in their market share due to bad press reports, or because they want to achieve a competitive advantage, companies adopt an ethical behaviour. However, companies' sustainable behaviour also depends on the consumer's interest in ethical products or companies. Such consumption has a political meaning: the rates of its growth, and its growing influence, are interpreted as signs of an increasing demand for sustainable products and will have consequences for normative decision-making.

On the other hand, finally, individual consumption is also targeted by public policies that expect consumers to behave in a more responsible and sustainable way (for instance, when consuming energy or water) and to try to influence them with prescriptive elements (e.g. taxes, information) while mobilising their 'citizen' identity as a way of justifying sustainable action-taking. Both as a consumer and as a subject of public policies, the individual can participate in sustainable development in its social and environmental aspects, but his or her responsibility is framed and limited by structural constraints. However, changes in consumption can and should be encouraged.

In this chapter, we are defending the idea according to which action should be meaningful for a social group if one wants to encourage the implementation of individual sustainable behaviour in the household (1) and that behavioural changes in the household also depend on social changes (2).

In order to implement sustainable behaviour in the household, individual action should be meaningful for a social group

Conventional economic models are based on the consumer's rationality and on the actor's equal position with other economic actors in the exchange process. It is far from reflecting the reality of life. Many elements make the situation a very complex one in which the consumer's capacity for assessing the opportunity of a change is difficult. First of all, information is not perfectly distributed. For example, labels that carry a sustainable quality (FSC, Max Havelaar, Organic) have low awareness levels and the consumer does not know much about the criteria that they guarantee (CRIOC, 2004). Second, the sociology of consumption has shown that the consumer's capacity for change is limited by many elements among which are the socio-material collective systems to which the consumer is connected (Otnes, 1988), the cultural system (Douglas, 1983), the social relationships (Bourdieu, 1979) and the importance of forgotten aspects such as comfort, cleanliness or convenience (Shove, 2001).[1] All these elements interfere with what may be understood as the consumer's rationality.

We would like to emphasise also the importance of another element of (un)sustainable consumption: the consumer's perception of a large-scale threat by opposition to an environmental risk that can be controlled or insured Luhmann (1993).

The risk issue is often forgotten in consumption studies, with the exception of the financial risks linked to mortgage and environmental risks (e.g. floods) that can be covered by an insurance contract. However, the risk issue in consumption is also a matter of safety, health and quality of life. To this extent, a high level of anxiety is created by high environmental risks (e.g. nuclear, greenhouse effect). Such anxiety has been well analysed by Giddens and Beck (Beck, 1987; Giddens, 1994), who have underlined its paralysing effect: individual action is meaningless in a world threatened by high environmental risks. This is as much linked to a perception of powerlessness in the face of the risk as to the risk itself (Scherer, 2004). In such a background, only activists or some opinion leaders have the will to adopt more sustainable behaviours.

However, recent work has shown that another figure of the sustainable consumer appears: what Dobré calls 'ordinary resistants' (Dobré, 2002) and Scherer 'domestic activists' (Scherer, 2004): those people who use their domestic behaviour as a political manifesto for the environment and as a way of taking action against environmental risks. In other words, they are the people who act – without repressing their anxiety – at an individual level while avoiding joining organised collective action. These types of ordinary resistants are also found in alternative consumption networks and in Fair Trade (see Pirotte,

Chapter 9, this volume). Anyway, domestic or political activists both share the idea that their personal action is meaningful for a social group. Therefore, what is at stake for individual sustainable action-taking is the meaning of action.

However, the issue for policy-makers is also to explore whether it is possible to expand the amount of people who would be interested in sustainable consumption beyond activists by integrating risk and solidarity values in the utilitarian assessment that underlies consumption behaviour (see 'Mobilising solidarity within a consumption logic' below).

Our conviction is that this can only be achieved if individuals are considered not as consumers but as citizens. This means going further than a simple appeal to moral values and towards a reconsideration of the structure of the consumers' 'and of the citizens' interpretative frames. We suggest that the citizen/ consumer logic (i.e. the famous *consum'actors*) cannot be fully mobilised within the framework of conventional economic concepts.

Mobilising solidarity within a consumption logic

Grin and Van de Graaf (Grin & Van de Graaf, 1999) analysed 'the way people formalize actions and meanings by referring to their interpretative frames'. This varies according to the situation in which people are and, during action, mobilises various interpretative frames. In what they call the *'first order'*, people mobilise their definition of problems and solutions and they assess these solutions according to their specific assessing schemes. For example, a consumer uses his or her experience and knowledge to consider a purchase (this stage can include the meaning of an artefact). He or she then determines what may be found on the supply side for answering his or her need and assesses them according to his or her evaluation scheme, which often includes a cost–benefit analysis. This may lead to a reconsideration of action. Such interpretative order is at stake most of the time.

However, Grin and Van de Graaf highlight a 'second order', described as normative and empirical theories and 'ontological' normative preferences. It represents the roots of action and is always present but seldom mobilised. When detailing the interpretative frame of a consumer, Grin and Van de Graaf suggest that, opposite to the conventional understanding that puts price rationality as underlying consumption, consumers' ontological preferences are based on their favourite state of well-being and comfort (Table 13.1). In other words, consumers are never going to buy or use something against their preferences in terms of well-being and comfort; the price issue will be considered only if these preferences are fulfilled.

This understanding has been very efficient in explaining why many people do not consume more sustainable products. Let us take the example of an efficient light bulb: even if the normative and empirical theories that underlie action are made up of environmental considerations, consumers are going to assess the solution represented by efficient light bulbs in terms of comfort and convenience before any other considerations. If the lighting quality is not what

Table 13.1 Interpretative frame of a consumer

First order	Evaluations of the pros and cons of the various domestic actions	Includes a cost–benefit analysis
	Definition of the problem at stake and of the meaning of the solutions	The problem definition can include the meaning of an artefact which is used in domestic tasks (or in the household)
Second order	Normative and empirical theories and beliefs	Expected social order
	Ontological preferences	Favourite state of well-being and comfort

Source: Grin and Van de Graaf (1999).

is expected (some people find it too feeble) such a lack of comfort is enough to prevent people from buying the efficient light bulb.

In other words, the environmental benefit of sustainable consumption is always going to be balanced by individual norms of comfort and convenience. It is all the more so when individual action is environmentally efficient only if taken up by many others (e.g. saving energy to counterbalance the greenhouse effect). In that case, individual action is easily meaningless (Beck, 1992) unless supported and meaningful for a collective (Scherer, 2004).

Hence, in our opinion, the failure of many projects aimed at implanting sustainable behaviours is due to their neglect of the collective meaning of individual action (and consumption). The issue at stake, then, is to consider the interpretative frame that could be efficiently mobilised for consuming in a more sustainable way, hence allowing the consumers' ontological preferences in terms of comfort to be overcome by ontological preferences that would allow the integration of social concerns. We suggest that referring to citizenship logic is a possible answer.

Mobilising solidarity within a citizenship logic

For implementing sustainable consumption among people who are not activists, many policies refer to the consumer as a citizen without questioning the types of links that exist between these two identities. Therefore, it is worth studying the quality of these links and considering whether solidarity is as naturally embodied in consumption logic as the words citizen/consumer seem to suggest.

As observation shows (Scherer, 2004), French individuals have a very strict understanding of the notion of 'citizenship' and of the actions they can take by mobilising this identity. They are restricted to the tasks that are expected of them by local institutions or by their neighbours (waste separation, avoiding littering the streets). In other words, they correspond to a social expectation that is:

- Expressed by social actors at all levels such as national and local authorities, local enterprises, but also neighbours (who blame you if you don't sort out your waste). To this extent, social expectations also correspond to a social control.
- Organised particularly with the help of artefacts (such as specific separation bins) that are displayed for helping the implementation of action. The meaning of individual action is given by its importance for the social group.

Such a form of socialisation carries the meaning of action to such an extent that action can turn into a habit without reference to its sustainable character anymore. People do it because they have to.[2] The social meaning of action corresponds to the ontological basis of action.

Policies aimed at implementing sustainable action within the household have the objective of helping sustainable action turn into a general habit. It is, for example, what waste separation policies want to achieve, as waste separation depends on the citizen's behaviour (Rumpala, 1999). When observing sustainable actions which are widely socialised in France, it was possible to describe the interpretative frame that corresponds to a citizen's logic more precisely. For example, we observed the following.

Fairness and equity are essential elements in the underlying normative and empiric theories of people's interpretative frames. Warlop *et al.* have observed the same element for Belgian households (2001). In other words, citizenship action is accepted if it mobilises the collective body involved in action with equity.

Most of the time, the definition of the environmental problem is imposed by the institutions. There will be much to say about such a one-sided way of considering collective public action. However, the meaning of the solutions is assessed according to very pragmatic criteria (e.g. frequency of collection, odours).

It is therefore possible to suggest a structure for the interpretative frame of a 'citizen' according to the previous criteria. This is summarised in Table 13.2.

Within that frame, individual solidarity consumption is only meaningful if it is so for the collective body to which the individual belongs.

This is not the case for many voluntary sustainable actions. For example, changing one's transportation mode is not coherent with people's interpretative frame as citizens, since the social meaning of such a move is not recognised by other social actors. Why use public transport when public authorities keep on building car-parks in the city centre? While many environmental policies expect citizens to include the 'moral' aspect of pro-environmental behaviour in their rational cost–benefit analysis because they can get positive benefits (including financial ones) out of behaviour change, such a conception is not coherent with the consumers' interpretative frame, which is anchored in comfort and convenience (the bus journey takes a longer time) or with the interpretative frame of a citizen anchored in the social meaning of action (none of the other social actors care anyway).

Table 13.2 Interpretative frame of a citizen for socialised sustainable action-taking

First order	Evaluations of the pros and cons of the various domestic actions	Causal assessment of solutions in terms of efficiency
	Definition of the problem	Imposed by the institutions
Second order	Normative and empirical theories	Fairness, equity
	Ontological preferences	Social meaning of the artefact or action

Source: Scherer (2004).

Against this background, the organisation of collective action for sustainable consuming takes its full importance as the key for implementing individual consumption changes. The social movement theory can help us understand what could influence this organisation.

Changes in individual consumption depend on the organisation of collective action

The individual does not exist outside a network of social relationships and society is the dynamic system made up of these relationships (Van Campenhoudt, 2001). Therefore, analysing change means analysing a system of concrete relationships and understanding its dynamic. Social change results from the behaviour of individuals who interact. More precisely, conflicts of interests allow power relationships and the implementation of a change process. They are canalised by a spatio-temporal system that provides margins of action to both the individuals and the organisations.

The process of change

According to the actionalist theory, action or actors' systems are defined according to intentions, cultural orientations and social relationships. A society cannot be reduced to economic determinisms or to functionalist patterns; it always tends to modify them or to go beyond them. In this theory, historicity is the action taken by society on its social and cultural practices in order to achieve a change. It combines three components: the mode of knowledge acquisition, the economic activity and the cultural model. The lower layers of the model are made up of social organisations and of the political system. They both interfere in the action system.

Social relationships try to master historicity. The social class that possesses the capital has the leadership. The popular class is dominated and opposed to the leaders. It struggles by proposing an alternative social model such as in the nineteenth century, when labour consciousness was high, unified and showed solidarity. Therefore in those examples, social relationships play a key role in the production of society. They induce change and social structuring.

To analyse change, a theory (i.e. a group of hypotheses that can be confronted with empirical observations) is needed. The social movement theory, which is also known as the actionalist theory, is made up of the convergence of three principles and three action levels (Touraine, 1978). An actor who is able to acquire a proud consciousness (principle of identity) when confronting an opponent (principle of opposition) behaves or fights for a general orientation of society and for the control of the whole social development (principle of totality). Three levels must be articulated: 'micro' for organisations and individuals; 'meso' for politics and institutions; 'macro' for development models. The movements that want to change the development model without anchoring themselves into a concrete organisational network and without relying on policy actions (designed for modifying the institutional framework) are utopia for Touraine. At the opposite end, when institutional and organisational work keeps being centred on internal challenges, it cannot contribute to the modification of both the development model and the use of resources.

Against the background of claims for a more sustainable development, such an approach has become a very complex one. The rise of the anti-globalisation movement is a phenomenon that must be taken into account when analysing social change. Whether this movement and others are able to generate a large social mobilisation that will be able to affect the whole society is still an open question.

Nowadays, values and health, ecological and ethical images appear in production modes and merchandising. This is not the result of sudden guilt but an effect of the numerous information and awareness-raising campaigns organised by groups of ecologists and consumerists.

By contrast with market consumerism, which targets prices and the material aspects of quality, the 'social consumerism' movement (born in Europe at the end of the 1980s) has raised awareness about the social condition of workers in the textile industry. Ethical concerns are therefore part of the deregulation and globalisation trends and challenge the supremacy of economic values, particularly within the World Trade Organisation.

Managing the change

> If collective action is such an important issue for our societies, it is because it is not a natural phenomenon. It is a social construction that raises concerns and whose conditions of birth and life are to be explained.
>
> (Crozier and Friedberg, 1977)

Change, within these approaches, is seen as a learning and collective creation process through which actors elaborate new ways of cooperation and conflict resolution, hence allowing the existence of new objectives for action, new priorities and new tools. If the system (and its constraints) determines the background of action and the actors' resources, it is also influenced by their calculations and manipulations. Change results both from the uncertainty game organised by activist movements and from the organisation of collective action as a model for action that associates the principles of identity, opposition and totality that characterise the actionalist theory.

Controlling uncertainty and managing change are the challenges brought by new social relationships. They can influence consumption greatly. However, we are going to examine if the actionalist theory remains perfectly appropriate to analyse the current process of social change.

The game of uncertainty

Responsibility about prevention and precaution should be shared between companies and consumers according to the preventive action and precaution principles that are at the heart of the European consumption and environmental policies.

European policy considers environmental and health risks for consumers from the early stages of the product's development to their final destination as waste (from cradle to grave). If potential harms for health and the environment should be corrected at their very source, then responsibility for their treatment cannot be transferred to the collectivity, the taxpayers and the environment. Against this background, dialogue between the political world and the world of consumption for action are of the utmost importance.

In the consumption field, there are many uncertainties linked to the increased importance of the concept of risk and of the structuring of ethics within companies. Because those risks can potentially harm both the environment and the consumer's interests, they are at stake in negotiation processes to the extent that the actors who are able to control uncertainty have a power advantage. It is to this extent that uncertainties can be real vectors of change, much more than through a simple assessment of the progress that has been made to control them and to increase their social acceptability. For example, the organisations involved in the social side of sustainable development have a great influence on information display. Most information comes from surveys, studies and descriptive analysis rather than from the assessment of companies' progress by means of indexes. Such a situation questions the role of technical and scientific information in the structuring of the consumer's reactions when confronted with risk issues and in the structuring of safety and quality demands.

The organisation of collective action

When referring to the interpretative frames quoted in the previous section, we observe that the meaning of action is unclear for everybody but the activists, and that giving back a meaning to action is only possible if collective action is organised. Actually, when taking action and trying to understand its meaning, the individual mobilises different identities that correspond to different interpretative frames. As long as the interpretative frame mobilised by the individual is that of a consumer, he or she will have little reason to participate in the solidarity economy as, above all, he or she will look for convenience, comfort and utility. In the best case scenario, he or she will consider solidarity criteria afterwards as the icing on the cake. Therefore, one of the aims of sustainable consumption policies should be to help the consumer change his or her interpretative frame in order to understand solidarity consumption.[3]

As mentioned previously, field researches held in companies and among consumer groups (Scherer, 2004) show that pro-environmental or solidarity action makes sense when it mobilises a citizen's interpretative frame, i.e. when it is organised not only with artefact and information but also when the collective body acknowledges the importance and the meaning of individual action. These characteristics may be found in the Belgian or Dutch eco-teams' experiences that aim at promoting sustainable behaviour in households. To the contrary, the isolation of the individual and the lack of involvement of local institutional actors prevent individual action-taking because they make it 'meaningless'.

This raises the issue of the role of institutions in the organisation of action. While the general trend is one of general institutional withdrawal, increasing the institutional role seems a precondition for the participation of individuals in solidarity economy. If we refer to the importance of the debate on risks, one of the goals of public action could be the organisation of collective action centred on this issue.

The socialisation of individual behaviour and its institutional challenges are not the only components of collective action: power relationships must also be taken into account. The rise of both new unconventional markets and lasting differentiation sources for the consumers show the potential strength of collective movements.

On the institutional side, the Organisation for Economic Development, United Nations, European Union and the International Labour Organisation have all established normative principles on environmental and social regulation. Such norms, or 'hypernorms', are closely related to public interests. Thus the enterprises are called to operate lines of action which are desirable in terms of values for our society. Various groups of stakeholders are stimulating the companies to perform in a socially responsible manner. As a rule they have the duty to meet stakeholder norms by developing desirable business behaviour (Maignan and Ferrell, 2004).

The stakeholders may be regrouped into four main categories: *organisational* (employees, customers, shareholders, suppliers, investors); *community* (local residents, specific interest groups); *regulatory* (municipalities, policy-makers), and *media stakeholders*. They all, directly or indirectly, affect or are affected by the firm's activities. The conflicting relations between enterprises and stakeholders are the driving forces bringing about new strategies of power and changes in society. What is questioned is how to analyse and manage the situations of potential conflicts.

The stakeholders (we mean especially specific groups such as consumer organisations, environment defence organisations and trade unions) put pressure on business and contribute to modify organisational processes within the companies. Three approaches are relevant to relate such an evolution:

1   The social performance management system is based on the integration of new practices depicted in the management literature as recommended drivers for the anticipation of future business challenges (Vlasselaer, 1997). The vision of long-term objectives, the strategies and programmes of actions, the monitoring of the whole value chain, the relaying of social or sustainability reports in addition to financial reports, and the review every three or five years, will not be efficient if they are not viewed as a central part of the corporate mission and the strategic plan of the firm. All these activities are focused on dialogue with the stakeholders. What is important is to intervene in the issues which are considered major by a given company. The company accounts for, and balances, the stakeholder interests as key value drivers.

2   The stakeholders are also the primary targets for business communication. Consumers are aware of the importance of buying more products developed in a sustainable fashion, but they do not convert their positive attitudes into purchasing decisions for different reasons. At the moment, the price of the product is considered as the main factor of choice, and the products are not easily accessible and visible. In addition, consumers do not trust the sincerity of the claims the companies are making and the proliferation of marketing terms creates consumer confusion.

Many studies show that what the consumers are demanding the most is credible information. Labelling programmes are drivers of consistent communication. Concerning ethical performance, credibility is of key importance for consumers because of their difficulty in evaluating ethical or immaterial values for themselves. Consumers form their opinion from the information to which they have access. This asymmetry should affect the nature of the information a company chooses to provide. The labelling programmes are a means to substantiate the claims and to make immaterial values more tangible even if in reality it is difficult for consumers to understand the information (see Rousseau and Bontinckx, Chapter 6, this volume).

Aside from labelling, there are of course other specific methods for companies to communicate superior performance (e.g. advertising, direct marketing, reporting). However, according to the signalling theory developed at the beginning of the 1970s (Spence, 1974), verbal communication or advertising, for example, may be regarded as a bad message (or bad signal) because it is available to all firms and is not able to distinguish competitive advantage in these contexts. On the contrary, official labels are connected to product strategy with strict norms and have the potential to create improvements in companies that extend far beyond benefits such as public image. For companies, the labelling programmes can promote:

- significant partnership with organisations (and the gain of increased stakeholder support);
- opportunities for strategic learning (e.g. life-cycle analysis, green technologies, reporting);
- the building of resources and competencies which enable the companies to stay ahead of imitative competitors.

The ability to learn faster and to learn more than competitors has become the new battleground for companies. In the short term, benchmarking will make it crucial to communicate the nature and the reason of competitive advantage. Consumers are given new sources of information. Several organisations make ethical ratings of the main manufacturing groups and the consumer organisations have adopted ethical items into their comparative tests.

3   Conflict management and game of uncertainty have evolved definitively and have given rise to new expressions of power in the relationship between social organisations and enterprises.

When interests derive explicitly from broader values and ideologies, the conflicts are more difficult to resolve than when interests are separated from values. Today, organisations distance themselves from ideological considerations and are ready to compromise more on their positions. This results in a new model of negotiating behaviour. This model is based on an integrative social contract and is governed by responsiveness in close relationships instead of reciprocity in bargaining relationships. Various indicators reflect such an evolution: joint interests versus one's own interests, self-monitoring of agreements versus regulation, relational concerns versus demands for concessions. The opponent is perceived as fair and trustworthy. This model impacts positively upon the process and outcome of negotiations as well as the effective implementation of agreements. However, there is a risk of subordination of interest groups due to the asymmetrical powers. On the other hand, minor failures can have significant effects on companies' reputation and value. Demands for transparency, benchmarking and watching systems are probably the new driving forces for balancing the relationship.

The European Commission and a number of players (e.g. large industrial and manufacturing groups, NGOs, trade unions, certifiers) now see the movement of corporate social responsibility (CSR) as the basis for the rules that must prevail in worldwide industrial and business trade by the end of this decade.

Increasing numbers of companies are devoting resources to showing commitment, asserting a certain ethic and adopting socially responsible behaviour. In only a few years, businesses have seen the formation of many social observation agencies (such as Vigeo – formerly ARESE – a company based in France, and *Sustainable Asset Management* – SAM – a company in Switzerland), the development of socially responsible investment and the emergence of behaviour aimed at putting pressure on companies deemed irresponsible. Finally, the legal framework is evolving with, for example, the introduction of laws aimed at making it compulsory to publish information of interest to society in the annual report.

However, we are sceptical about the effectiveness of these tools for imposing changes with regard to crucial stakes of the present unsustainable world.

Although managers generally have a good understanding of the roles they play in defining procedures for managing the environment and social issues, they do not always think about acting transparently and communicating with the stakeholders. Yet this is key to the success of the system. Resistance to various forms of external control is usually an additional restraint.

It is no longer simply a matter of shouldering responsibility *after the event* but of adopting a behaviour that anticipates the consequences of actions carried out and of the underlying causes. Despite the attention paid by companies to the concept of CSR, the nature of that concept and its implications remain vague. Actions generally focus on the specific aspects of CSR, such as sponsorship or protecting the environment, and do not give an integrated picture of it. How should internal conflicts within the same company be reconciled: Is a company with a high level of employment at the price of a high level of pollution more or less responsible than one that uses state-of-the-art technology which is much less polluting but creates unemployment? How should its short-term performance be assessed when the environmental and social initiatives are part of long-term strategic objectives?

Conclusions

Mobilising solidarity within a consumption frame and its utilitarian background is bound to fail because it does not fit with the interpretative frame that is mobilised by the consumer. However, when mobilised within the individual's interpretative frame as a citizen, it gives a meaning to action that can overtake consumption preferences and lead to changes in behaviour, on condition that individual action is also meaningful for the collective body (company, institutions, local groups) to which the individual refers. Therefore, mobilising citizen logic cannot be achieved within the frame of conventional economics and its cost–benefit analysis. It necessitates the introduction of other criteria such as equity between people and a collective endorsement of change through social norms.

These aspects are at the heart of production and consumption changes towards more sustainability, and public institutions have a specific responsibility for and a specific capacity of action in this process. Against this background, a specific focus should be on the influence of the individual's construction of the environmental risk on the process of change.

Giddens states that modernity is a source of anxiety because the individual is confronted with multiple risks and has to trust the capacity of expert systems to manage them. Therefore, the individual must decide who he or she is going to trust, hence constantly building his or her own identity as well. When trust is lacking, people withdraw into themselves and sometimes deny the existence of the risk itself. As defined by Giddens, risk and trust are major elements that could be the basis of the construction of collective action. To this extent, institutional actors have an important role to play because, when they reduce their action to the regulation of the economy in its conventional understanding, they limit the power game that allows individuals to build their identities and the companies to change their behaviour.

For companies, solidarity is then reduced to an ethic of responsibility that goes along with their duty of integration in the society within which they operate. By answering to market demands, companies redefine their citizenship and the criteria of their economic efficiency. Considering the social pressures, such an environmental and social involvement is a necessity. More and more companies consider it as a new management model that can provide economic performances with a capacity to last and to resist crises. Within that frame, controlling uncertainty, promoting a trustworthy image, defending one's reputation, access to new markets, process management, and the evolution of internal organisational processes are the challenges brought by new social relationships. They can influence consumption greatly. However, if uncertainties linked to the risk issue and to the ethical structuring within companies have modified social relationships, they have also modified the social actors' capacity for inducing change. It is true that the dialogue between companies and social organisations has taken on a higher importance. This is certainly relevant in the current process of social change. But we are far from the logics that found the actionalist theory.

Power relationships change according to the institutional game. Against this background, companies are both powerful and vulnerable. Citizens lack references that could allow them to consume better and to ask for credibility that companies cannot provide. The trust they demand could be a basis on which organisations and public authorities could build a process of change towards a more sustainable economy. However, such a process appears to be very different from what the actionalist theory analyses to the extent that its capacity of inducing a social change on the wide scale can be questioned. The actionalist theory shows how a group of individuals with common interests can gain a position strong enough to be successful in its demands when faced with an opponent. Three conditions are necessary in this way (the three principles on which the theory is based): the global will to change world and society (principle of totality), a collective identity around the same objectives of change (principle of identity), and to stand against an indicated opponent (principle of opposition). Nowadays, we may wonder if the theory remains perfectly appropriate to analyse the current process of social change. The requirements of interest groups or social organisations are most often limited to particular and concrete claims and the political will to change the consumption society is not really present.

Notes

1 For a summary of the various approaches of the sociology of consumption, see Jackson and Michaelis (2003).
2 There is no financial penalty in France for not complying with the expected behaviour.
3 Referring to consumers as potential collective actors raises the issue of the social basis of collective action. According to a recent survey, it seems that the lower classes of Belgian society are more interested in ethical products than are the upper classes (CRIOC, 2004). This may be a potential basis for collective action to which little attention is paid. The same observation may be made for other potential social bases such as territorial ones.

14 Is large-scale Fair Trade possible?

Ronan Le Velly

Introduction

More than a billion euros' worth of Fair Trade products were sold in the world in 2005. As a result of this relative commercial success, attention is now being paid to its potential as a way to combat poverty and contribute to sustainable development, despite the fact that Fair Trade has existed in most European countries for some three decades. The aim of this chapter is to show that this means more than just setting figures to the market shares that Fair Trade could achieve. Such a study is useful, but tends to mask the fact that Fair Trade is not a homogeneous reality. When we ask if large-scale Fair Trade is possible, we are actually trying to describe what type of Fair Trade may be envisioned on which scale. The question is not so much about the possibilities of outlets as about the changes in the trade relationship that the rising sales induce.

These questions come out of a field study I conducted of the main French promoters of Fair Trade, namely Artisans du Monde and Max Havelaar (see Box 14.1). In 2005, the network of 'Artisans du Monde' shops posted a turnover of 10 million euros (this was twice the 2000 turnover), while a total of 120 million euros' worth of products bearing the Max Havelaar label were sold in France (this was twenty times as much as in 2000). Such advances cannot be explained without referring to the efforts that have been made over the past fifteen years. The 'Artisans du Monde' network has increased and modernised its points of sale, and the quality of the foodstuffs and craftwork that they sell has improved considerably. Finally, under the impetus of Max Havelaar, Fair Trade products have come on to supermarket shelves. We are far from Fair Trade such as it existed in France up until the early 1990s, that is, focused on tiny off-putting shops that a few rare consumers patronised as deliberate acts of commitment, buying coffee that was 'disgusting' but 'Sandinista'.[1]

Yet for all that, Fair Trade advocates do not have an unequivocal inter-pretation of these developments. The activists are obviously overjoyed by the rising sales, which, as they point out, make it possible to reach a larger number of producers. However, on the other hand, the policies implemented to achieve these results can generate some unpleasant feelings. Many mention their fear of 'selling their souls to the Devil' and, as a result of the compromises they must

Box 14.1   The Artisans du Monde network and Max Havelaar France system

The first Artisans du Monde shop opened in Paris in 1974, and the Artisans du Monde network is currently the largest French network of shops specialised in selling Fair Trade products (150 points of sale in 2005). The shops are run by volunteers, often aided by a part-time or full-time salaried employee. For the most part, they are not supplied directly with craftwork and foodstuffs from the Third World producers' groups whose commodities they sell, but order their supplies from specialised Fair Trade importers. Their primary supplier is Solidar'Monde, which was created at the initiative of the Artisans du Monde Federation in 1984.

Max Havelaar France, which was founded in 1992 following the example of the Dutch initiative of the same name, is a certifying body. In placing its logo on products, it guarantees that these goods meet a set of formal criteria concerning production conditions and purchasing from the producers. In 2005, there were more than 100 registered traders in France (Alter Eco, Lobodis, Malongo, Solidar'Monde and so on). They are the ones who do the work of importing and processing the commodities and searching for sales outlets. The commodities bearing the Max Havelaar label (e.g. coffee, tea, bananas) are then sold to consumers in all sorts of sales circuits, but mostly in hyper- and supermarkets.

The French organisations have been working in close cooperation with their foreign counterparts for the past ten years or so to harmonise the criteria for defining Fair Trade and to pool certain operations. Solidar'Monde is a member of the European Fair Trade Association (EFTA), which is a union of Fair Trade organisations (Gepa in Germany, Traidcraft in the United Kingdom and so on) that coordinates its members' activities when it comes to selecting and monitoring producers' groups. Similarly, Fairtrade Labelling Organisations International (FLO) sets the standards of the labelled system. FLO is an association of a score of national initiatives (Max Havelaar in France and the Netherlands, Transfair in Germany, Fairtrade in Great Britain and the United States, and so on). As a member of this framework, Max Havelaar France entrusts the operations of certifying producers' groups to FLO's auditors.

make, 'ending up as simple merchants'. These worries are largely the result of the hybrid, if not contradictory, nature of the Fair Trade plan itself. Fair Trade advocates want to build a different model of trade to oppose what they perceive to be the malfunctioning of 'conventional trade'. However, as soon as they also use certain capitalist economic gears to increase their sales, the question of the Fair Trade graft's being accepted or rejected arises.

One finding of this study is that this positioning does indeed generate contradictions. This chapter therefore looks into these tensions and how Artisans du Monde and Max Havelaar France's activists see them. It does not concern either the producers or the consumers directly, but takes as its starting point observation of the work to build trade between these two opposite ends of the production–consumption chain that is both opposed to and within the market.[2] We shall see first how difficult it is to offer products that meet the quality demands of the greatest number while working with 'small producers'. Then, in the same way, we shall see that fair prices are not set completely without reference to conventional market prices. Finally, the last two sections will be devoted to the centralisation of imports in the Artisans du Monde network and the strategy of delegating commercial operations that is specific to the Max Havelaar system. These two situations will lead us to consider Fair Trade as a continuum of positioning assigning more or less weight to the desire to oppose or to participate fully in the market.

Which producers for which products?

Studying the characteristics of the producers' groups targeted by Fair Trade is a way to approach the difficulty of being both against and in the market. This first section will be based on a reading of the principles and standards defined within EFTA and FLO (see Box 14.1), and personal observations of and interviews with French Fair Trade activists. This will enable us to grasp who the 'small producers' with whom these activists wish to work are and to understand why it is so painful to demand that they produce 'marketable products'.

'Small producers'

The term 'small producers' has many meanings, but they are all aimed at building a type of trade that differs from conventional trade. First of all, the activists speak very often of 'cooperatives of small producers' to assert the importance of the principles of democratic organisation and the observance of human rights in the workplace, even though all the groups are far from having this legal status. The cooperative ideal is thought of in opposition to the image of the large corporation employing a wage-earning workforce and is assumed to be the epitome of domination and exploitation. Next, the term 'small producers' refers to the goal of preserving cultural traditions. Artisans du Monde's activists want the products that are sold in their shops to be made according to models and techniques that are faithful to their country of origin's traditions. In so doing, they wish to showcase the richness of their partners' skills and know-how. But, even more important, through this criterion they wish to show that they are not involved in a dominant relationship, one of sending out orders that the craftsman fills. From this perspective, working with small-scale producers seems to be a guarantee of true craftsmanship and cottage-industry work involving little mechanisation, standardisation and acculturation of the goods. Finally, and most important, through this term the Fair Trade organisations are targeting 'marginalised small producers'. Fair Trade strives to give organisations that which conventional trade does not: because of their small size, low investment level, handicaps or discrimination, small producers are described as having no access to the conventional market or failing to get satisfactory payment for their work. The conventional market is seen as functioning to the detriment of the 'disadvantaged producers who do not have the means to make it on their own and finally live at the mercy of large companies'.[3] Working with small producers without wielding the market power that this procures is yet another way to assert Fair Trade's difference.

'Marketable products'

Since the early 1990s, Fair Trade advocates in France and the other European countries alike have no longer contented themselves with a symbolic denunciatory function. They want to expand market outlets beyond dyed-in-the-wool activists in order to support the producers' development fully. To do this, efforts to 'professionalise' operations were made first in the North (locating the shops in more central shopping areas, training the volunteers in sales techniques, hiring employees, selling through mass distribution, advertising and so on).[4] However, this demand for professionalisation also concerns the South. EFTA's principles state very explicitly that producers' groups must 'seek to produce a marketable product'.[5] There, too, the aim is to reach a larger number of consumers, but in the activists' view, the aim is also to sell quality products that establish a relationship of dignity, rather than charity, between producers and consumers.

So, several questions in the assessment questionnaire that EFTA's members use are aimed at the 'small producers' that we have just been discussing. These are questions about working conditions (health and safety), procedures of democratic representation, degree of marginality, community development projects carried out and so on. However, the information that is requested is also aimed at determining whether the imported goods meet European technical standards and consumer tastes (e.g. quality approach, innovation, design) and whether the producers have sufficient production and export capacities. The first consequence of this approach is that the craftwork increasingly undergoes cultural adaptation. Even though some people continue to regret this, even though they speak about 'team work' and 'local designers' to defuse criticism, it is currently acknowledged that the products are made to meet Western consumers' tastes without necessarily adhering to the strict criterion of cultural authenticity (Littrell and Dickson, 1999; Grimes and Milgram, 2000). This also affects the nature of the producers who are selected. For example, in 2002 the importer Solidar'Monde (see Box 14.1) asked Artisans du Monde Federation's Board of Directors to choose between two Burkina Faso bronze casters' organisations. It submitted the following opinion to help inform the Board's choice:

The Toure Issaka group has the advantage of being a group of authentic traditional bronze casters, which is not the case of Zod Neere . . . [but] one of the reasons that make us more inclined to choose Zod Neere is that . . . Toure Issaka seems to work as if coping with emergencies and in a situation of extreme [financial] precariousness from which it is impossible for them to extricate themselves without external support inside the country. It is hard for us to see how this group could evolve and shake off its precariousness, even though, of course, having work for part of the year would probably be of considerable benefit to them in the short run. . . . Zod Neere, in contrast, seems to think about and have an approach aimed at development. . . . And lastly, it is probably more reliable and more sustainable.

(Solidar'Monde, 2002)[6]

The Federation heeded this opinion and chose Zod Neere. It preferred an organisation with community development plans and market capacities that were described as high to another group that was more culturally authentic and marginal. In the past, opposite choices have been made and Solidar'Monde continues to work with some highly marginalised informal structures (especially a few partners who have been on the scene since its creation in the 1980s). However, today, when a new group is selected, awareness of the constraints associated with market participation leads to renouncing some of the principles that are specific to working with 'small producers'.

In the same vein, FLO's standards state, 'the producers must have access to the logistical, administrative and technical means to bring a quality product to the market'.[7] For example, to apply for inclusion in the coffee register, a group must first of all fill out a questionnaire that specifies the means of communication and transport at its disposal and send in a sample of the coffee that it could export. The importers and industrialists that sell large volumes of coffee to supermarkets and/or food chains such as Starbucks refuse to have poor-quality products and demand regular deliveries (Argenti, 2004). Consequently, the producers' groups in FLO's register cannot be the most marginalised. This trend is reflected first of all in their strong geographical concentration. In October 2003, 33 per cent of the suppliers (all products together) were in Central America, 26 per cent in South America, 17 per cent in Asia, 14 per cent in Africa, and 10 per cent in the Caribbean. Mexico alone had 16 per cent of the organisations on the register.[8] These figures may be explained by the history of Fair Trade, but they also stem from the Mexican producers' better production and shipping capacities compared with Black African or Haitian producers. Next, several field studies have confirmed that purchases within each country are clustered around the most developed organisations on the register and the least structured groups fail to win Fair Trade orders (Eberhart and Chaveau, 2002; Shreck, 2002; Murray *et al.*, 2003).

How to set the producers' remunerations?

The ambition of boosting sales, even if that means being more competitive, is a source of injunctions that contradict the assertion of trade turned towards marginalised 'small producers' who uphold cultural traditions. This contradiction between the will to participate actively in the market and the ideal of opposing the characteristics that are attributed to the conventional market is also clearly visible in the ways the prices that are paid to the producers are set.

'Fair prices'

A poster used by 'Artisans du Monde' in the early 1990s bore the following succinct text: 'Du café: juste un commerce ou un commerce plus juste?' ('Coffee: just trade or trade with justice?'). In just a few words, the network asserted the assumed unfairness of conventional trade and the need to create an alternative. Speaking of fair prices implies a principle of setting the prices paid to producers that escapes the impersonal confrontation of supply and demand. It is in precisely this perspective that FLO strives to establish minimum prices for most of its products that are disconnected from market forces. The formula for calculating the minimum price that is being discussed within FLO should thus allow for production costs, the costs that are attached to converting to meet Fair Trade criteria (for example, the reorganisation of work), what is deemed to be a reasonable profit margin, and a bonus enabling the producers' groups to improve their production capacities and living conditions. On the other hand, it should not include the global production volume or consumer market prices, which are parameters associated with the way the conventional market functions. The aim is to assert a different way of framing the picture – a 'reverse mechanism', according to its proponents – that marks the specificity of Fair Trade.

Market prices

This anti-market frame, however, is never totally closed to market influences (Callon, 1998). For example, when world coffee prices hit rock bottom in late 2001 the Fair Trade coffee sold in German and Swedish supermarkets cost twice as much as conventional coffee. While consumers agreed that the 'Fair Trade' label justified a higher price, such a difference made the product practically unsellable. Given this context, FLO mulled over the possibility of lowering the minimum price of coffee, and rumours that it would drop to US\$1.00 per pound (instead of US\$1.21) circulated throughout 2002. The idea of creating a fund for marketing operations was also considered as a more indirect way of coping with the crunch. Finally, the fair price of coffee remained unchanged and the creation of a marketing fund was postponed. It is true that the invoicing of certification visits that was decided in 2003 already places a financial burden on the producers. In addition, the gap between the minimum price and commodity market price for coffee has narrowed greatly over the past few years.

The tension concerns the level of the minimum price, but it can also concern the very existence of a minimum price. Thus not all FLO standards impose a minimum price. The most noteworthy exception is tea, for which the producer and importer freely negotiate a market price to which a 'development premium' in the amount set by FLO is added. The negotiated price is supposed to 'at least cover [the] cost of production',[9] but nothing sets it formally. When the rice standards were drawn up in 2001, there was a debate within FLO between the advocates of a 'market approach' and those of a 'minimum price approach'. The former argued their case based upon the risk of completely disconnecting purchasing prices from conventional market conditions and the latter upon the danger that freely negotiated prices might not cover production costs. The upshot of these discussions was the drafting in 2003 of rice standards that involved the payment of a 10 to 12 per cent premium above the market price and then, one year later, the drafting of new standards that this time set minimum prices.

There are also no minimum prices for craftwork imported by EFTA members. The problem for these goods hinges on the difficulty of establishing a nomenclature for the items being traded. The purchase prices are set case by case and take account not only of the producers' production costs and needs, but also of the prices that the items could fetch in the North. Some items that are deemed too expensive thus cannot be bought, possibly despite the value of their makers' development projects. For other goods, a round of bargaining is launched to achieve lower rates. Consequently, it happens that the purchase prices paid by Fair Trade importers are the same as those paid by conventional trade buyers. Many activists find this difficult to swallow, given that they are used to associating a fair price with one that is necessarily above the market price or hold that the importer should never challenge the producer's asking price. Yet this does not mean in itself that the prices paid do not enable the producers and their families to earn a decent living or to 'live in dignity', to quote the campaign slogan. These activists' reactions attest above all to the discomfort generated by Fair Trade's current positioning in the scheme of things. Being opposed to the market is not enough to be protected from market forces.

Direct or intermediated relationship?

Working with 'small producers' and setting 'fair prices' calls for knowledge of the producers' groups' situations and taking this into account in establishing trade relations. However, the objective of personalising relations does not stop there. Fair Trade activists want to establish direct links and 'interknowledge' between producers and consumers.

'The idea of the invisible hand has given way to the idea of working hand in hand'[10]

Fair Trade is opposed to the presence of local middlemen or moneylenders ('loan sharks' or 'usurers') who use their positions of strength to impose their conditions on 'small producers'. According to a deeply rooted stereotype among Fair Trade activists, such parties are 'coyotes', 'parasites' that get rich at the workers' expense. Fair Trade is then described, for example, in Artisans du Monde's teaching kits, as including fewer middlemen than conventional trade. Moreover, EFTA's principles and FLO's standards alike stipulate that Fair Trade importers must pay for a part of their orders in advance so that the producers do not have to borrow funds at exorbitant rates.

The goal of a direct, personalised relationship also means that the importers are involved in a lasting relationship with the producers. In a stationary market such as that of craftwork, such a stand corresponds to a principle of non-competition. When the Artisans du Monde network already has a partner for a specific type of crafted article, it abstains from working with another group that would compete with its current partner. The activists consider this an indispensable rule, once again to distinguish Fair Trade from the conventional market. To use the famous terminology, voice and loyalty must be preferred to defection (Hirschman, 1970). The lack of a personalised, lasting commitment would lead the discussion back solely to the matter of products and prices and result in behaviour that would be incompatible with the producers' development.

Finally, Artisans du Monde and Max Havelaar's activists want to have 'inter-knowledge' relationships with producers. They want to know how the producers work, exchange information about their daily lives, and know what their plans, projects and difficulties are. While this link with the 'producers behind the products' is important for the activists themselves, it is also important because it must be extended to consumers. Even if the physical distance between the parties does not change, Michael Goodman explains, it is nevertheless possible to reduce the feeling of distance and otherness such that 'the well-off "us" (consumers) and poorer "them" (producers) becomes a "we" (participants in the same network)' (Goodman, 2004: 907). In Artisans du Monde's shops, this link is forged by displaying panels depicting the groups or their locations on a large world map. Information, snapshots and testimonials are also printed on the foodstuffs' packaging or on small cards attached to the craft items that are bought. Finally, the salespeople see it as their duty to discuss development issues with their customers and fill them in on the product's 'story'.

Ultimately, the dream Fair Trade chain is a transparent relationship of reciprocal knowledge and respect between producers and consumers in a market devoid of merchants. Of course, Fair Trade chains actually operate in less black-and-white and, above all, more variable, ways.

Intermediation and impersonalisation in the Artisans du Monde chain

In the early 1980s, Artisans du Monde had some twenty shops in France and no centralised importing structure. The shops placed their orders directly with the producers' groups that they had found through common acquaintances and then exchanged items among themselves in order to increase their ranges. Solidar'Monde was created in 1984 for the purpose of importing foodstuffs. Since the latter came from other European Fair Trade agents, this did not trigger any major discussion within the movement. However, debate quickly arose around the possibility of extending this centralisation to craft items in order to rationalise ordering and warehousing. It nevertheless took two general assemblies of the Artisans du Monde Federation to reach agreement on the principle of centralising craft imports. The main reasons for the members' reservations had to do with the disappearance of the direct, personalised link that would result from Solidar'Monde's intermediation. This argument explains why even today some Artisans du Monde groups continue to import items directly as a sideline. Many of the volunteers in the particular group I observed feel that such relations are vital. For example, the group buys items from a centre for physically disabled children in India that was created by a volunteer's uncle. At each meeting this volunteer reads out letters recounting life at the centre. These moments have all the characteristics of information about friends whose lives the group has been following for years, whose successes are reasons for rejoicing (for example, when the centre's children are at the head of their classes in the local school) and whose problems are announced cautiously ('I'm afraid that the news is not good . . .').

Solidar'Monde tries to maintain such ties with the producers by transmitting written information about them to Artisans du Monde's activists and by organising visits for producers' representatives, but the relations that are instituted are less regular and intense than those that the decentralised import schemes made possible. Yet for all that, there are no plans to return to the old system. Centralised purchasing management enables the network to rely on a larger number of producers' groups, greatly facilitates stock and delivery management, and makes the creation of a wide, coordinated and frequently renewed product range possible. In a context in which active participation in the market is advocated (selling more to support more producers) and competition between Fair Trade importers is fierce, centralising imports appears to be inevitable for the Artisans du Monde network.

This confirms the opposition between the injunction to participate in the market and that of opposing the market. However, we are also starting to understand better that the arbitration between these two injunctions can take several forms. The shops' direct imports and centralised importing thus appear to be two ways of structuring the chain that assigns relative weights to the desire to be commercially effective and the desire to create a 'different kind of trade'. One volunteer with a long history in the Artisans du Monde movement expresses this positioning difference perfectly as she stresses not only the degree of personalisation but also the type of producer involved and the shops' price-setting procedures:

> [When it comes to] Solidar'Monde, it has to stand on its own two feet; things must be balanced; it is not charity work. And it is true that sometimes we deplored their refusal to work with small cooperatives, with small groups . . . as I objected, 'That is our main reason for being!' Yes, it's wonderful, for example, to import embroidery from Palestine, but if it doesn't sell, whom does it help? What good is it? So, we were truly harassed, we had requests from Palestinian women, it was horrible, truly horrible. . . . We were directly in touch with these Palestinian women, and they wrote us heart-rending letters, and then their embroidery didn't sell, because a selling price had to be set, and when it comes to prices, Solidar'Monde is a hard bargainer: If it doesn't sell, they don't take it . . . Artisans du Monde aims to be more structured. [Silence] In the beginning, Artisans du Monde was like that, somewhat, very strong on relationships, and as it became more structured we lost a little of that. That is why the long-standing members here are highly committed to maintaining relations with direct partners, because some of that comes through. We get news from them; they write to us, we wonder what is happening when we get no news from them.
>
> (Artisans du Monde, 2003)[11]

Controlling the chain or delegating business activities?

The idea that I ultimately wish to defend is that of a continuum of possible positionings assigning different relative weights to the imperatives of effective business and opposition to the market. Seen from this standpoint, the labelled chain set up under FLO appears to be located on a third level of market participation after direct and integrated chains. This induces some significant perverse effects regarding the goal of creating a 'different type of trade'.

Delegation as a sales-boosting strategy

The difference between the integrated chain (Artisans du Monde) and the labelled chain (Max Havelaar) lies in the classic alternative of doing something and having it done. The integrated chain's organisations run their commercial operations themselves. The labelled chain, in contrast, is marked by twofold delegation, that of selling and importing. Several types of agent are involved in the sales end (e.g. restaurants, food chains, catering, mail-order houses), but the main targets are hyper- and supermarkets. Max Havelaar France's employees and volunteers are very quick to criticise the way the major distribution chains work, but in a country where, as they systematically point out, 80 per cent of consumer goods are sold via these circuits, refusing to take part in the mass distribution system amounts to action throttling the producers' development. In other words, whereas the Artisans du Monde network has fewer than 150 shops across the entire country, the 10,000 POS in France that sell labelled products give each and every resident of France access to Fair Trade. So, Max Havelaar's Dutch founders, Nico Roozen and Frans Vanderhoff (2002), explain that they would have liked to have imported and processed the products for mass distribution themselves, but had to give up such ideals due to the lack of financial means and marketing expertise of the international solidarity organisations that supported their project. The labelling strategy was thus a way of recognising that large-scale business activity was not within the scope of the organisations spawned by solidarity associations and it would be better to delegate such tasks to external agents.

The perverse effects of delegating activities

The labelling–delegating strategy aims to change the scale of Fair Trade. Moreover, there is no doubt that it is perfectly effective on this score. Six to seven times more Fair Trade products were sold in supermarkets than in specialised shops in 2004 in France alone (this same ratio is borne out across Europe as a whole – see Krier, 2005). However, delegating activities also has perverse effects on the ambition of creating a type of trade that differs from conventional market activities.

First of all, delegating a task increases its impersonality. Compared with Solidar'Monde's employees, Max Havelaar France's employees have little reason to be in touch with the producers and are less able to provide information about them. Next, the centralisation of monitoring and support functions for the producers' organisations under the FLO system has increased the northern activists' distance from the Southern producers even more (Murray *et al.*, 2003). Finally, the creation of links between producers and consumers is weakened by selling through supermarkets. It is difficult to post more information than what can fit on the package and impossible to be physically present in each and every supermarket to tell shoppers about the producers' living conditions.

This loss of contact goes hand in hand with a loss of control. In integrated chains, Fair Trade associations control the commercial activities that are carried out. In labelled chains, they merely check that practices comply with pre-established standards. FLO does not choose directly the organisations that benefit from Fair Trade. By setting up a register of groupings that meet its criteria, it merely draws up a shortlist of sources from which the registered traders select their suppliers. Importers in the labelled chain are thus perfectly free to prefer to work with already well-structured groupings rather than with 'marginalised small producers'. They can also prefer private plantations to 'small farmers' cooperatives'. Most of the FLO registers do not propose such a choice, but when they do, as in the case of bananas or tea, competition between private plantations and cooperatives generally ends up favouring the former, given their ability to deliver more constant quality and at more regular intervals

(Shreck, 2002). The difference between the integrated and labelled chains in this respect is important. If Artisans du Monde's activists wish to, they can force Solidar'Monde, of which they are shareholders and the main customer, to work more with groups that fit the 'small farmer' or 'craftsman' image. Max Havelaar's activists cannot impose such demands on the commercial operators unless they push through a difficult change in their standards.

Comprehension is dawning that delegating these tasks entails a change in the market's structure and creates conditions for competition between the producers' groups. Being included in the producers' register proves that an organisation meets Fair Trade standards but does not automatically result in purchases. Forty per cent of the organisations in the coffee register have never had a single order under Fair Trade terms (Eberhart and Chaveau, 2002). All in all, only one-fifth of the tonnage of coffee produced under Fair Trade conditions is bought at the minimum price set by FLO. The rest is sold at global market prices or slightly above the market price when the additional quality warrants it. The producers are thus objectively in a situation of competition and Fair Trade importers have the power to bargain in their favour. However, things do not stop there. The delegation strategy also leads to granting the label to all traders who meet FLO's trade standards and then letting them look for distributors on their own. Some twenty coffee roasters were offering Fair Trade coffee to supermarkets in France in early 2005. (The number of Fair Trade suppliers of other commodities, such as chocolate and bananas, is more limited for the time being.) Given the very high concentration of the French supermarket sector, the roasters are under great pressure. As Marie-Christine Renard points out so well: 'paradoxically, while this network was intended to avoid mechanisms for competition, they begin to appear' (Renard, 1999: 498). As a result, we may add, whereas the aim of Fair Trade is to cancel out downstream market power, the labelling–delegation strategy ends up restoring this power.

This pressure, which comes to bear first on the registered traders and then on the producers, can have effects that are contrary to the plan of building alternative trade. Coffee farmers agree to sell the non-Fair Trade parts of their harvests at below-market prices in exchange for increases in the volumes sold at the fair minimum price. This practice, which FLO calls 'bonded contracts', is an indirect way of reducing the fair minimum price. Similarly, the partial prefinancing of harvests is not systematic. Here, too, there is not really any fraud, for, according to FLO standards, this prefinancing is done 'on the request of the seller'.[12] The importer's bargaining position can then be one of forgoing prefinancing in exchange for a larger order. Such malfunctioning is more improbable in integrated chains. Solidar'Monde and its European counterparts cannot use a non-Fair Trade part of an order in bargaining over the Fair Trade purchasing conditions. Next, the shops for them are relatively captive outlets that do not demand high profit margins. Finally, the commitment to a long-term business relationship that is so strongly asserted in the integrated chains tends to reduce the possibility of such manipulations. In contrast, the FLO standards are rather undemanding in this regard. While, in principle, 'buyers and sellers will procure to establish a long-term and stable relationship',[13] the formal requirements do not exceed one season. Labelled Fair Trade is a still young and fragile initiative and imposing too demanding standards would run the risk of dissuading commercial players from taking part in it.

This does not mean that the levels of FLO's standards are fundamentally low. In reality, they are definitely much higher, especially when it comes to the fair price and collective representation criteria, than those of competing labels such as Bioéquitable, Rainforest Alliance and Utz Kapeh. The Bioéquitable label has won over only a few small companies in France so far, but in the United States the Rainforest Alliance's criteria have attracted the food giants Kraft Foods, Procter & Gamble and Chiquita. Similarly, Sara Lee is working with Utz Kapeh to certify part of its Douwe Egberts coffee in the Netherlands and Belgium. For FLO's members, who are financed by the licence fees that their registered traders pay, these situations mean losses to make up and a risk of bankruptcy. Once again, this situation is the result of a delegation strategy that leads to being dependent on commercial operators' participation. If FLO does not want to lower its standards, it can only bank on its image and customer knowledge and recognition. This is a difficult wager, given that the latter are even less used to looking at what lies behind a label than behind a product. While Fair Trade contests commodity fetishism, one of the hobbles on its development is precisely the predominance of this fetishism in consumers' behaviour (Hudson and Hudson, 2003).

Conclusions

This chapter does not evaluate the match between Fair Trade principles and practices. It shows, rather, that the presence of contradictory principles makes such an assessment extremely subjective. Boosting sales is now an integral part of the Fair Trade plan, on a par with constructing a different form of trade. For example, saying that Solidar'Monde's intermediation harms the personalisation of trade relationships is not enough to condemn this intermediation, given the additional number of outlets it creates for Third World producers' groups. Similarly, while the labelling–delegating strategy induces regrettable malfunctioning, it remains a way to boost the market for the producers' outputs considerably. Several types of Fair Trade are possible, and direct imports by shops, the integrated chain and the labelled chain are in this regard three stages on a continuum in which practices are turned more and more towards market participation and guided less and less by opposition to the market. The question that can be raised is that of the limits of such a continuum. Is it still possible to talk about Fair Trade when such criteria of opposition to the market as a minimum price and collective representation have been dropped, as the Bioéquitable, Rainforest Alliance, and Utz Kapeh labels have done?

More generally, this analysis of Fair Trade updates and refines Max Weber's (1978) pessimistic finding that it is impossible to rebuild within the capitalist system an economy governed by substantive rationality that is mindful of people and concerned about moral, religious, political and aesthetic imperatives.[14] Several recent economic sociology studies suggest that Max Weber was doubtless wrong to consider the capitalist market order's constraints to be a homogeneous, omnipotent force. Market transactions in capitalist economies can accommodate a variety of justifications meeting different principles of fairness without one's necessarily assuming that formal, calculating, self-interested rationality is at work (Boltanski and Thévenot, [1991] 2006). Moreover, there are numerous situations of markets in which the written rules, cultural representations and/or balances of power interfere deeply with the impersonal mechanism of matching supply and demand that Weber described (Fligstein, 2001; Zelizer, 2005). In showing the social underpinnings of the various Fair Trade chains' construction and the variety of ethical principles that preside over their development, the work summarised in this chapter confirms these analyses' relevance (for more in the same vein, see Raynolds, 2002). Despite all that, this work also cautions against focusing only on the social conditions of market building. The constraints linked to participating in the capitalist market order of which Weber (1978) spoke must be treated with discernment, but must not be taken out of the analysis. Like Andrew Sayer (2001), I believe that one cannot understand how markets work by postulating the existence of a market force that is independent from social considerations or concentrating exclusively on the variety of contexts in which exchanges take shape. The social construction of market transactions and constraints of the market order are mutually determined and must be examined together. So, when Artisans du Monde and Max Havelaar's activists call for fairer and less anonymous relations with the producers of the South, this of course refers to a cultural representation of the ravages wreaked by the conventional market, but also results from a very real trend in the capitalist world that they feel all the more strongly for being caught in its grip.

Notes

1 See Gautier Pirotte (Chapter 9, this volume) for a description of Oxfam World Shops' (*Magasins du Monde/Wereldwinkels'*) patrons in Belgium.
2 I translate the relatively neutral French term of '*filière*' as 'chain' and prefer not to use the expressions 'commodity chain', 'value chain' and 'supply chain', which have more specific theoretical meanings. See Raikes *et al.* (2000) and Raynolds (2002).
3 Max Havelaar France volunteer interviewed on 18 February 2004.
4 For a presentation of marketing techniques' application to Fair Trade, see De Pelsmacker *et al.* (Chapter 8, this volume).
5 EFTA, *Fair Trade Guidelines*, 1996.
6 Solidar'Monde, *Candidatures producteurs bronziers Burkina Faso*, 18 January 2002.
7 FLO, *Generic Fair Trade Standards for Small Farmers' Organisations*, January 2003.
8 Max Havelaar France, *Fair(e) Actualités*, January 2004.
9 FLO, *Fair Trade Standards for Tea*, October 2004.
10 Brid Bowen in EFTA (2001: 25).

11 Artisans du Monde volunteer interviewed on 5 February 2003.
12 FLO, *Fair Trade Standards for Coffee*, June 2004.
13 Ibid.
14 The following sections in the two volumes of *Economy and Society* raise these issues directly: 'Formal and substantive rationality of economic action', 'The disintegration of the household: the rise of the calculative spirit and the modern capitalist enterprise', 'Religious ethics and the world: economics', 'The market: its impersonality and ethic', and 'Hierocratic and charismatic ethics versus non-ethical capitalism' (Weber, 1978: 85–86, 375–381, 576–590, 635–640, 1185–1188).

15 Impact of Fair Trade in the South

An example from the Indian cotton sector

Isabelle Scherer-Haynes

Consumers choose goods according to qualities that are both imaginary and objective, but the way in which those goods are produced is not a concern for them so long as some basic regulations are followed; in terms of hygiene, for example. It is the contrary for several goods for which the conditions of production represent part of the quality that the consumer is looking for. Fair Trade products belong to this category of goods. According to Fair Trade (FT) standards (EFTA, 2001; IFAT, 2005), Fair Trade goods are paid at a fair price, i.e. a price that allows producers to cover their production costs and earn enough money to have a decent living; something that is not the case, according to Fair Trade actors, in the conventional trade system. But the aim of Fair Trade is also the empowerment and development of small producers of the South and contributing to achieving greater equity in international trade by offering small producers the opportunity to trade and by helping them to get organised 'as stakeholders in their own organisations'. In order to achieve that, they promote producers' organisations and the payment of a premium, added to the fair price, which is going to be used for community development and managed by those organisations.

While Fair Trade actors criticise the conventional international exchange, consumers are becoming politically involved when buying Fair Trade goods (Poncelet *et al.*, 2003) or are engaging in a charitable act. In other words, they become involved in what Goodman (2004) calls a transnational moral economy. A couple of authors have chosen to analyse Fair Trade by showing that it can be considered as a network where the relationship between the consumers and the producers – hidden in conventional trade – is reconstructed (Whatmore and Thorne, 1997; Mardsen, 2000; Raynolds, 2002). This is done thanks to a Latourian socio-technical device ('*dispositif*') which makes the figure of the producers present through personalisation and information about their work and life conditions (Goodman, 2005). More generally speaking, Fair Trade, as a quality network, connects specific producers' organisations (partners) to specific Fair Trade intermediate actors in the North and then to consumers.

However, the necessity of expanding markets to provide more sale opportunities for the small producers in the South has led FT organisations to enter mainstream markets. This movement brings into the network many new actants such as labelling organisations, supermarkets and also a new type of consumer who is less aware of the specificities of Fair Trade but with high expectations as to the respect of basic social rules (Haynes, forthcoming). Such a movement may be described as an expansion of the FT network in Europe which has consequences for the organisation and knowledge of FT organisations and FT producers in the South (see also Le Velly, Chapter 14, this volume).

Few papers (e.g. Hughes, 2000) take into account the notion of network for analysing the consequences of the entry of Fair Trade in conventional trade at the Southern producer and Southern FT organisation levels. The objective of this chapter is to contribute to this analysis through the description of the changes in knowledge and organisation in an Indian FT organisation. In doing so, we will refer to the concept of a socio-technical network. It is anchored in the Actant Network Theory (Latour, 2005) which observes the recomposition dynamics within a network and invites the researcher to pay symmetrical attention to both human and non-human actants and to the links they create between each other. This understanding of the network allows the identification of important actors which are not identified by the conventional analysis of commodity chains.[1]

It is also a help in underlying the importance of the mediations that create links between various actants and allow the functioning of the network, especially when it is destabilised and has therefore to recompose itself. To this extent, the 'enrolment' of the actors in the network (that is, giving them a specific task as essential elements of the network and increasing their involvement, hence providing meaning to action) is an important part of a network's construction and consolidation.

We also refer to Appaduraï (1986) who suggests that the social life of a good mobilises three types of knowledge: the knowledge necessary for its production, the knowledge necessary for its consumption and the knowledge that allows its circulation. Actually, Fair Trade mobilises these three types of knowledge. It organises an important transfer of knowledge from the North to the South in helping the circulation of goods (in terms of international trade rules, export organisation and marketing). In this way, it substitutes itself for the traditional intermediaries that control prices but it has to provide several facilities, such as an opportunity of market access to the producers.

It also plays a role in the knowledge needed for consumption as we indicated that the figure of the producers, which tends to be erased in mainstream trade, is highlighted in Fair Trade. Moreover, it interferes in the knowledge necessary for production by conveying environmental (integrated crop management) and social criteria. Finally, it tries to introduce a political knowledge in the management of the benefits by asking the producers to form an organisation that will function democratically.

The fieldwork for this research, which was financed by the Belgian Federal Science Policy, was held in 2005 in the Indian state of Gujarat, in the Katchchh area that is affected by an endemic drought, increased salinity of soils and where many catastrophic natural events have occurred in past years (see e.g.

Ramachandran and Saihjee, 2001; Deaton, 2003). Many farmers depend on cotton crops, the price of which varies a lot according to the international market and local conditions. For example, it went from 1,400 INR^2/40 kg in 2004 to 800 INR/40 kg for the winter 2004/2005 campaign. In this situation, the debts of farmers increase as they can no longer afford the pesticides and fertilisers they need. Poverty increases to the extent that some farmers who cannot pay their debts commit suicide (see e.g. Sharma, 2004).

Against this background we had the opportunity of working with company A,[3] an organisation that participates in the FT cotton network that links Belgian Fair Trade consumers with Indian cotton producers. We interviewed cotton farmers at home, members of the staff and the founding manager of company A. We also went to Mumbai, where the apparel workshops working at the end of the production line on company A's cotton are located. We interviewed the managers and, in one case, the staff without the presence of the manager. These interviews were completed in Gujarat by the women from Shrujan, another NGO embroidering cotton T-shirts for company A, and by interviews with importers and distributors of FT cotton products in Belgium.

First, we will describe the changes linked to the extension of the network to mainstream distributors which have new demands, particularly the traceability system that guarantees to the consumer the origin of a product. Second, we analyse the consequences of both the integration of transformation actors in the network and the normative process linked to the labelling scheme that is necessary to assert the 'fair' quality of cotton products in the mainstream market.

Changes linked to the demand of mainstream markets

Before 2005, FT labels for cotton did not exist and company A, as a partner of Oxfam World Shops in Belgium, could benefit from the positive image of Oxfam World Shops among consumers (SONECOM, 2005). However, entering mainstream markets meant obtaining a label that would certify the quality of the cotton. It implies traceability of the product in order to guarantee that it was not mixed with conventional cotton. Mainstream distributors also ask their providers to respect a certain amount of rules about the physical quality of the products which place new organisational constraints on the Southern producers. Both demands mobilise increased knowledge of the production and circulation of cotton products that are supported by the introduction of new actants into the network.

Changes in knowledge about traceability influence network organisation

The traceability system guarantees to the consumer the origin of the good he or she buys. Knowledge about traceability norms involved in Fair Trade has led to a specific organisation of cotton circulation thanks to a socio-technical 'device' that is implemented both at the farmers' and at the transformation levels.

At the production level

Company A had to acquire knowledge about the traceability issue and the way to implement its requirements in terms of crop isolation and documentation; i.e. it had to make the farmers understand the necessity of documenting their practices while translating that necessity into a proper organisation that the farmers would respect.

Materially, the necessity of isolation has led to the storage organisation of the cotton crop in separate rooms at the farmer's home, often in one of the farmer's living rooms, its collection using specific trucks (company A does not participate in the collection of conventional cotton), and its storage with the other crops in a specific warehouse.

Traceability also meant the implementation of a whole documentation system about cultural practices: as we mentioned, Fair Trade organisations commit themselves to the implementation of integrated crop management (ICM).[4] Although cotton farmers follow ICM practices, documenting their work is an issue that is still not very well understood by some of the farmers we talked to. 'I always have everything in my head, I know my land by heart, why bother writing?' The fact that properties are not very big (10 to 15 acres)[5] contributes to this question. The farmers do it because it is required but still don't understand its necessity, which raises questions about the degree and even the existence of their enrolment in the FT network. Actually it does not appear that the existence and principles of FT or the advantages provided by a traceability system for consumers have been explained to the farmers (nor were they understood by them). Most farmers did not know about FT. This suggests that the FT network can work without full enrolment of all the actors.

The books (that I could see) are kept in Gujarâti, which means that they can only be read by Indians unless FT actors from the North learn the language or company A translates the main information into English. If FT requirements are fulfilled, the existence of this situation underlines the fact that the relationships that exist between company A and Northern controllers are also based on trust.

At the transformation level

FT cotton transformation must also follow traceability rules, which asks for the enrolment of new actants in the FT network.

Traceability of the product means that FT cotton must be transformed separately. Company A had to take into account the material aspect of ginning and spinning; that is, ginning factories will not accept less than 50 tons of cotton and mills less than 10 tons or this would otherwise affect their profitability too much (the machines have to be stopped and cleaned between the processing of conventional and Fair Trade cotton). These quantities represent a technical barrier to alternative cotton production that has already been described by Mrill (2000). The impact of this constraint is that it has led to the organisation of a cotton bank where the cotton is stored until the minimum volume is reached. This bank has turned into a marketing tool as it is also of help in providing cotton to customers after the picking season. If the activities of company A expand, it may also allow the organisation to act as a trader on the international market of quality cotton, where it will be the one of the primary actors from the South.

Impacts of changes in knowledge about quality standards on the organisation of the network

Entering a Fair Trade network also means adopting conventional market rules even though Fair Trade criticises them. It is the paradox of Fair Trade, which is at the same time both inside and outside the market, respecting formal international market rules while breaching some of them in respect of the producers' work conditions and remuneration (Le Velly, 2004; see also Chapter 14, this volume).

To enter the international cotton trade, company A had to learn the rules of the cotton market, the international export standards and the quality standards which are defined according to American standards (see Box 15.1).

At the farm level, the cotton is submitted to the market's normative expectations in terms of quality for the compliance to which non-humans are important elements. For example, farmers had to learn to manage their crops with more care so as to avoid the inclusion of items such as cigarette butts, pieces of plastic and husks, which contribute to the deterioration in the quality of the industrial processing of the cotton (referred to by cotton specialists as a low 'contamination' level). To this extent, contamination is limited by the implementation of simple artefacts that are included in the socio-material device implemented with traceability: transportation of the crop is made with cotton bags instead of plastic ones to avoid polyester contamination. Farmers are

---

Box 15.1 Cotton international quality standards

The United States Cotton Futures Act of 1914 authorised the Department of Agriculture to establish physical standards as a means of determining colour grade, staple length and strength, and other qualities and properties. These standards were thereafter agreed upon and accepted by the leading European cotton associations and exchanges. They were accordingly termed and referred to as the 'Universal Standards for American Cotton'. These standards are commonly used by twenty-four signatory cotton associations representing twenty-one countries that have signed the Universal Cotton Standards Agreement and by twenty-five non-signatory countries. Whereas other countries like Benin started developing their own classification system, the USDA is still the most commonly used standard.

Source: CNUCED (2005)

enrolled in the quality network thanks to a simple educational tool: they are confronted with yarns that include plastic and can assess the amount of yarn that is wasted accordingly. These rejected samples help them to understand the consequences of their practices including the reasons why their cotton could be rejected. Such artefacts make the issue and its financial consequences understandable. Their existence contrasts with the absence of artefacts that would be aimed at supporting the enrolment of farmers in the FT network. Finally, handpicking was also a traditional practice that contributes to cotton quality, as it avoids the presence of husks in the crop.

Both farmers' and company A's efforts in avoiding contamination of the ICM cotton resulted in a quality level that was high enough to allow the registration of A's cotton fibres, which implied for company A the acquisition of other legal and technical competences. We note that the links between mainstream markets' and FT expectations is the same in Africa. When organising the Fair Trade cotton chain in West Africa, Max Havelaar France (January 2005) completed it with a Quality Chart that strongly resembled company A's organisational device.[6]

Company A was also aware that, after transformation, cotton products had to comply with conventional markets' expectations in terms of delivery time, quality and fashion. These demands are much more severe than those of Northern FT buyers who accept delays in delivery time and lower quality products much more easily (interview, FT organisation, 2004), all the more so as their traditional customers were ready to accept a lower quality level (IDEA Consult, 2002). This had specific impacts for the non-governmental organisations (NGOs) working with company A, particularly on the embroidering activity.

Impact of the requirements for physical quality

Like those in the farms, craftswomen who embroider cotton clothes, as well as sewers, have to understand the quality issue, which means acceptance of a specific material organisation device to reinforce quality. To avoid stains, women are asked to sew or embroider in specific rooms, sometimes in a specific workshop where children are not allowed; they must protect the floor or the table where they work with mats.

Impact of delivery requirements

Even though Fair Trade organisations are much more flexible and ready to accept delays in delivery or quality problems than conventional organisations, the limits of this acceptance are narrowing because Fair Trade organisations are facing sales difficulties in Europe. This can lead to complicated situations because, for example, hand-made production of what is often a complementary activity hardly fits into the delays that are required by Northern markets. For this reason, visits from Northern Fair Trade organisation professionals are expected, as it seems to be the only way of making the North understand the difficult conditions of production.

Impact of fashion and design constraints

When looking at the knowledge necessary for the circulation of goods, there are quality problems linked to the hand-made aspect, as this type of production does not fit with the market expectations in terms of product standardisation. Sometimes, this difficulty is linked to a specific technique, such as block printing (patterns cannot be reproduced with perfect homogeneity). Sometimes, the tension occurs between demand and India's culture. For example, some women do not understand the necessity of producing exactly the same piece all the time and cannot help but add their personal touch, all the more so that exact reproduction can bring bad luck. It is the way in which transformers consider their work and its importance in the organisation of their life that is at stake.

Knowledge about Northern norms in the fashion industry is also a problem. For example, sewers were talking about the difficulty of understanding new professional signs like patterns (which are said to be different from Indian ones) or vocabulary (for example, a pleat is often understood as a gather). Therefore the workers (tailors, sewers) are involved in a learning process. But it is not a one-way learning process. On their side, to help the products fit with European markets, some FT organisations in Europe or in the South have hired stylists and/or people responsible for production who can provide guidelines while understanding the production difficulties.

Some knowledge is considered important for fashion design, for example, knowledge of European garment cuts and European sizes (it is so different from Indian standards that the input of a Western stylist is considered compulsory) and knowledge about decorative patterns (surface styling). In that case, collaboration between Indian and European stylists is possible. It is even more the case for crafts.

Designers are also aware of the consumers' tastes and demands and are able to draw new product lines that can be sold more easily in Europe. Actually, even though FT consumption is often analysed as a consumption of commitment, we have observed (Haynes, forthcoming) that committed consumers also pay attention to intrinsic product quality such as the shape, the motif or the finish of clothes. However, according to designers and managers in both India and Belgium, it is also a question of designers in Europe being taught to work with cotton, something that design students do not know much about any more because most of their exercises are done with polyester or mixed fibres.

Summarising, integrating mainstream markets was possible thanks to a socio-technical device, the existence of which is not a guarantee of the enrolment of producers into the FT network. Furthermore, tensions between local culture and market expectation may be observed. However, new tensions and difficulties are linked to the respect for social standards in transformation, difficulties that are enhanced by the newly published FLO/Max Havelaar standards for Indian cotton that require changes which challenge company A's culture and organisation.

Transformation and certification mobilise new knowledge that challenges the organisation and functioning of the FT network

Cotton is specific to FT as it is one of the few FT agricultural products transformed in the South. This raises concerns about the social conditions of its transformation. Furthermore, in 2005, the arrival of two new actants – the FLO together with new standards for Indian FT cotton and its certifying organisation, Max Havelaar – also led to new transformations. These new actants create new relationships which lead to an extension of the FT network while reducing the space for negotiation that has allowed the coordination around FT norms between company A and Northern FT actors. It may have an impact on the organisation of company A and its survival within the FT network.

Matching the geographical extension of the chain with the enrolment of actors in the FT network

It was obvious for FT intermediate actors both in the North and in India that the social conditions of cotton transformation were important. Hence company A had to control the respect of social laws in the entire chain until the end product, i.e. in companies it does not control. As the apparel industry which transforms company A's cotton is located in Mumbai, this meant expanding the network geographically, which has an implication in terms of knowledge and organisation.

According to the unionists or middlemen interviewed in India, if ginning, spinning and weaving are done in factories that mostly respect Indian social laws, this is rarely the case in the apparel industry which is smaller, very flexible and sometimes even difficult to identify because many workers are home-based.

To avoid these problems, company A chose to work with workshops run by various NGOs. But the development of production (for example, a children's line) has led it to work with conventional workshops with different work and social practices. The controversy is that, even though work conditions are better in the selected workshops than in those of most of the apparel industry, these workshops pay, as independent workers, men and women who could be considered as their employees. Actually these men and women do not possess their own production tools (the sewing machines) and come every day to the workshop while respecting its schedules. The situation has positive financial benefits for the workshops: workers are paid on a piece-by-piece basis, which means that rejected pieces are not paid for. The cost for the workshop is less than the cost of wages. It is also a way of avoiding stronger legal social constraints linked to the employment of more than fifteen employees. The controversy is embodied in a 'work card', the delivery of which will make the workers acquire the status of employees. The issue is then to convince those actors to change their social rules and provide work cards, even though they cannot foresee the financial benefits of a change: company A is not their main business source. In other words, the challenge is to make these actors share Fair Trade values, to enrol them in the Fair Trade network while they already belong to the organisation of the network. Company A has chosen to enrol these actors through negotiation rather than force.

Company A's position is that the situation will improve over time and that convincing its contractors to change their behaviour is more efficient than imposing rules on them. 'We begin by trying to make them [apparel workshop owners] accept the elements that don't require major investments: the creation of toilets, cleanliness and so on. It is more difficult when you touch on the capital, as the first thing they ask is to guarantee the orders in the long run, but progress can be made and we are trying to implement a Memorandum of Understanding [MoU] with the transformers', says an A company manager. The Memorandum of Understanding is a written document with no legal status in which each part explains what it expects from the other. Observing that they were also used by some NGOs to define their relationships with the institutions, a hypothesis that we could not verify is that, in a country where oral commitment is binding when made among what we could call a 'community of trust' (that is, interlinked people or organisations sharing the same objectives), such tools could be enrolled when crossing the border between different communities of trust (such as the FT and apparel) is necessary.

Company A's position is coherent with Gandhi's ideas which constitute the philosophical background of its action,[7] particularly with the idea of non-violence:

> But it must be realised that reform cannot be rushed. If it is to be brought about by non-violent means, it can only be done by education of both the haves and have-nots. The former must be sure that they will never have force used against them. The have-nots must be educated to know that no one can really compel them to do anything against their will and that they can secure freedom by learning the art of non-violence. . . . An atmosphere of trust and mutual respect has to be established at the preliminary step. There can be no violent conflict between the classes and the masses.
>
> (Gandhi, quoted in Narayan, 1946)

It is also coherent with Gandhi's idea of independence and protection of local industries. According to Gandhi, only when production exceeds needs should the excess be sold and imports should be limited as much as possible in order to reduce India's dependence. In this respect, company A works with existing conventional workshops and, opposite to other Indian FT actors, refuses the idea of creating a new workshop that would respect exactly Northern expectations for social norms. 'India's capacity of cotton transformation is already three times bigger than its production, no further investment is needed', says company A's management. Actually, delocalisation exists within India and

Mumbai's workshops are in crisis because confection workshops have appeared in other states.[8] The idea of developing India's self-support for food production is thanks to an agriculture based on these principles participating in the 'green revolution'. Actually, company A activities began as a pesticides dealer but changed its analysis of independence and worked on the implementation of integrated crop management techniques.

A good way to help farmers get out of poverty was also to help them end their dependence on local traders and price variations, and to secure the access to the market by themselves. This is why company A took the opportunity of developing the cultivation of Fair Trade cotton with the support of Oxfam Belgium.

Until recently, company A's preference for negotiation was respected by Northern FT importers, who kept pressure on the organisation when discussing social rules in workshops but accepted its decision. This space for negotiation may disappear.

Impacts of the Fair Trade normative process

Max Havelaar's requirements for the certification of Indian cotton impose social standards for the companies involved in the transformation of cotton, while the FLO standard imposes the creation of formal producers' organisations. The implementation of these norms participates in a shift from negotiation to control which tends to reduce the space for negotiation that exists between company A and its Northern partners.

The social conditions of cotton transformation

In the absence of social standards for transformation, organisations that wished to increase their added value by guaranteeing the social conditions of cotton transformation have the choice between different criteria that are 'private', i.e. not legally recognised – also being based on the International Labour Organisation (ILO) standards – mainly the Clean Clothes Campaign code of conduct, the SA 8,000 standard and, in the Netherlands and Belgium, the code of the Fair Wear Foundation.[9] These standards are both expensive for Southern organisations that must bear their costs and sometimes not very appropriate: they are designed for factories, not for small workshops. Therefore, many organisations have chosen to avoid them and to directly negotiate the application of ILO rules with their Northern FT partners. The new FLO standards allow the continuation of this situation because they fail to mention social standards for transformation. This situation may change as, in the press release that goes with the introduction of FT certified cotton products, Max Havelaar mentions that it will control the respect of the ILO conventions (Max Havelaar France, January 2005). Such a strategy is compatible with the condition of transformation implemented in the African network certified by Max Havelaar where most of the transformation is made in Europe, and with some Indian networks where European industries and supermarkets work with Indian partners and impose their own rules (e.g. BioRe for Maikaal).[10] On the opposite side it will be difficult for company A, which is not linked specifically to a major European importer, to comply immediately with those standards at least in the absence of larger orders for cotton that would help negotiation with the transformers. This situation raises two issues.

First, if larger orders are given thanks to the help of Max Havelaar which can find new clients, it means that many more cotton producers should be enrolled in the network. Company A is willing to take this step so long as purchase is guaranteed to the producers in the long run, something that Max Havelaar cannot do, as it proposes a list of FT producers from which Northern importers choose their partners and do not guarantee any purchases. Actually, company A has refused to drastically increase its production with no regard to the efforts asked of the producers, hence fulfilling its role of protection of the small producers.

Second, the choice for the promoting bodies is either to keep on working with the existing workshops or to create their own production line from scratch. If this last solution guarantees control of the respect of social standards in transformation, it forces the promoting bodies or the producer to acquire a specific knowledge, a new skill: the management of an apparel production unit. Acquiring this knowledge is a real challenge, and it might be easier if the promoting bodies are already positioned at this point of the chain: this is the case, for example, of Rajlakshmi Ltd. This company, which buys fair cotton from independent producers and transforms it, was considering implementing the Clean Clothes Campaign code of conduct in the factory it owns. Creating one's production line also requires financial assets. Rajlakshmi Ltd talks about hundreds of thousands of dollars for building a new factory that will be totally compatible with European standards. This is not within the reach of every producer or promotion body in the Fair Trade network. It means that the FT network cannot be analysed as one with an equal distribution of capacity and power.

The cost of certification

Furthermore, certification used to be free for producers and paid for by Northern actors. This is no longer the case, as Max Havelaar invoices fees to the producers' groups. They are calculated in thousands of dollars. Initial inspection costs for example vary from 2000 to 2500 dollars (FLO, 2004) and are in addition to other certification costs (e.g. SA8000, organic). Therefore, the capacity of Southern FT actors to comply with Max Havelaar requirements varies according to their position in the FT cotton network and according to their financial assets. Small farmers' organisations might not be able to pay the required fees, which leads to an exclusion process.

Against this background, company A is not as well equipped in terms of knowledge and finance as some other organisations. Furthermore, certification by Max Havelaar means that company A will have to adopt a formal control of its partners that is not coherent with its negotiation values. Such a dilemma may also be encountered when studying another norm imposed by the FLO: creating farmers' organisations.

The creation of formal producer organisations

According to FLO standards:

> The continual demonstration of the efforts of this organisation (the promoting body) to support the producers to reach a level of organisation that can comply over time with one of the existing Generic Standards is a condition to apply to Fairtrade. The standard is specifically to encourage the formation of autonomous producer organisations where they do not exist.
>
> (Fair Trade Labelling Organisation, 2005)

Complying with those standards implies a knowledge transfer from the promoting bodies to the producers and the mastering of new knowledge for the promoting bodies.

The ongoing necessity of increasing markets makes company A ready to organise the formation of autonomous producer organisations that would be able to discuss the use of the Fair Trade premium. It is not difficult because it already organises meetings with the farmers to discuss prices and agricultural issues. In the long run, the creation of such an organisation implies that farmers acquire some knowledge of management and marketing issues. A way of complying with FLO standards could be the opening up of the company's capital to the farmers. According to importers and company A's staff, such an organisation was considered at the creation of company A but abandoned on the grounds that farmers wanted the families traditionally responsible for trade to take responsibility for this aspect of the business. The responsibility of these families is also coherent with Gandhi's notion of 'trusteeship'. This approach insists on the moral responsibility of owners towards their business and employees. It promotes the idea of public good: owners are asked to consider themselves responsible for the public good which they must manage in the interests of all (Jaffrelot, 2005). Actually, the family who contributed to the financing of company A at its creation was in that situation. Even if company A is now depending on bank loans and not on this family's financial inputs, it is still very influential.

When interviewing farmers, the issue of getting shares or not was considered differently. First of all, as mentioned above, most of the farmers had never heard of Fair Trade, one of the reasons being that they joined company A with the aim of cultivating organic cotton. Actually, company A went from ICM to organic cotton cultivation[11] because it was another opportunity to gain access to international markets while reducing farmers' dependence on chemical inputs. All farmers working with company A in the FT network are now organic farmers. However, company A cannot afford to pay a fairer price to the producers (8 to 10 per cent above the market price) plus an organic premium. When a farmer enters the network, his production is paid at the FT price during the conversion years (three years on average) in order to compensate for productivity losses. When certified organic, the price supplement for production is converted into an organic premium but stays at the same level. For this reason, even though they benefit from company A's education programme and marketing, most farmers perceive company A as an actor in the organic network.

Second, many farmers wanted to stay independent and be able to sell their crop to whomever they pleased. The idea of creating a formal organisation did not fit with these expectations. The larger farmers expressed a concern for marketing in the sense that they were expecting new organic buyers to come to Gujarat, hoping that competition for organic cotton would increase the organic premium. None of them, perhaps due to lack of support, foresaw getting involved in the marketing for a community of farmers. To this extent, the farmers' expectations may not be exactly in tune with FLO's.

In the long run, a choice will have to be made between the need for a Max Havelaar certification and company A's present culture and form of organisation. Keeping its culture and its way of doing things is therefore challenged by the demands of FLO and Max Havelaar.

Conclusions

In India, entering a FT network has led to the implementation of a socio-technical device that conveys knowledge about norms: the norms that are necessary for the circulation of cotton products. They are coupled with the quality norms that correspond to the international trade in cotton. To this extent, this situation reflects the position of Fair Trade which is altogether inside and outside the market.

The FT network works thanks to a strong link between FT intermediate organisations in India and in Europe while the enrolment of other actors in the network is weak or difficult to implement. At the farmers' level, enrolment in the FT network does not always exist, while enrolment in quality is achieved thanks to the circulation of specific artefacts. By contrast, this emphasises the need for artefacts embodying knowledge about Fair Trade and about the links that are established with Northern consumers. But the Fair Trade cotton organisations in India also have to enrol transformers as, according to FT objectives, the added value of transformation should be kept in the South. This is a fairly new situation for Fair Trade and a difficult one for organisations that are more used to dealing with farmers' problems than with the respect of social laws in an industry that they cannot master. This extension of the network does not correspond with the enrolment of all its components especially in the apparel industry, in which controversy is embodied in the identity work cards that should be given to each worker.

Confronted with those difficulties, Southern FT organisations are not equal: those such as company A, which are close to the cotton producers' work, encounter difficulties in mastering what is going on at the end of the channel, while it is easier for other organisations which are positioned as transformers. Furthermore, Northern and Southern FT organisations used to be linked by negotiation processes that allowed an interpretation of FT norms by Southern actors according to their own culture and values. Such a situation is now threatened by the incorporation of FT norms into an official code of conduct created by FLO and controlled by Max Havelaar which carries a Northern understanding of the norms. On another level, cotton products are not always as standardised as expected because such an expectation is contrary to the transformers' techniques and culture.

Perhaps because it is not carried by any actants or embodied in any artefact exported outside of India, the specificity of these producers' culture and the importance of Ghandian thoughts for company A is not well known in Europe. Among all the Western people we interviewed, only some of those who made the trip to Gujarat were aware of it. However, it may provide a way of understanding the strategy of these actors. The question is then whether Northern Fair Trade actors who are used to transferring knowledge to the South should organise the circulation of information in the reverse in order to better take into account the cultural specificities of their Southern partners. It is not easy so long as actors from the South are not themselves involved in a reflexive process that makes them think about their specificity within the Northern conceptions. Furthermore, they are in a weaker position than the buyers, who can easily shift to products from other origins. On the other hand, lack of knowledge about the culture of Southern FT actors can generate lack of understanding of their behaviour and perhaps a lack of trust in their capacity and need to enrol in the FT network. The balance between laissez-faire and understanding of Southern cultures is difficult. However, respecting local culture is part of the objectives of Fair Trade and, above all, it is the work and life of many poor cotton producers that will be affected in the long run if some Northern Fair Trade organisations stop buying their products.

Notes

1 This expression is a translation of the French word 'filières'. For a comparative analysis of 'filières' and GCC, see Raikes *et al.* (2000).
2 1 INR (Indian rupee) = €0.01769.
3 At the time of publishing we have not received the company's authorisation to publish its name. Nevertheless, we take this opportunity to thank the management and employees of all the organisations that we could visit for their help and openness. They gave us their time while there is so much to do. They are taken here as an example of the difficulties of adapting to Northern norms that are common to other Southern FT organisations (Haynes, forthcoming).
4 Integrated crop management is aimed at reducing the amount of chemical fertilisers and pesticides. On a world scale, cotton farming uses 10 per cent of pesticides and

22.5 per cent of all insecticides applied in agriculture for 2.5 per cent of agricultural land (Ton, 2002).

5 1 acre = 0.40 hectares.

6 It asks for an early collection in order to allow traceability, the progressive replacement of all the polyester bags used for collection by cotton bags, improvement of storage and transport of cotton seed, and improvement of packing conditions for cotton fibre, the separation of cotton according to quality items in order to keep the best quality in reserve for Fair Trade clients and specific management of collection and sowing.

7 Company A's upper management was educated in Gandhian schools and company A farmers are the cotton providers of Gandhi's ashram.

8 We were able to get an idea of this impact when visiting, in Mumbai, a group of women who were at home finishing off mainstream confection work (getting rid of threads, sewing buttons and so on). They all found it increasingly difficult to get work; many of their husbands once had jobs in the apparel industry and were now unemployed.

9 These standards are available on those organisations' websites: http://www.cleanclothes.org/codes/ccccode.htm, http://www.cepaa.org/SA8000/SA8000.htm, http://www.fairwear.nl

10 http://www.biore.ch.

11 As recommended as well by FT standards.

16 Conclusions

The future of sustainable consumption

Paul-Marie Boulanger and Edwin Zaccaï

At the end of this collection of analyses of different aspects of sustainable consumption, we will attempt to go back to certain essential aspects of the role and motivations of consumers that are part of this approach. We will then deal with the difficulty of finding objective criteria to define it and finally we will raise certain points on which practitioners as well as researchers work today and that determine its reality and future.

Individual or collective action, consumer or citizen, private or public action

The role of consumers is a central theme of this book. Some of the contributions to it allow us to think about these roles, in particular in their individual or collective scope.

Many of the texts assembled herein (Fraselle and Scherer-Haynes, Le Velly) place the consumer in opposition to the citizen and suggest that if the consumer can act with 'responsibility', that is to say, subordinate his consumption to ethical or moral imperatives, such behaviour will rapidly meet the limits of its efficacy because of its individual nature. Only political action would be able to see the advent of an authentically sustainable consumption, which implies transformation of the consumer into a citizen. We find this very clearly expressed by Fraselle and Scherer-Haynes, who write: 'Our conviction is that this [sustainable consumption] can only be achieved if the individuals are considered not as consumers but as citizens. This means going further than a simple appeal to moral values.' To a certain extent, this position would confirm the opinion of Rousseau and Bontinckx's interviewees according to which the responsibility to protect the environment rests with the public authorities and the producers, and not really with the consumers.

Others, on the contrary, such as Pirotte (quoting on this topic Chessel and Cochoy, 2004) or Dobré, consider that responsible consumption includes from the start an intrinsic political dimension. With regard to this, Dobré speaks of 'ordinary resistance'. Pirotte gives a much more voluntarist vision when he writes: 'Buying organic, ethical or equitable products . . . is therefore a profoundly political act'; he even goes so far as to talk about an absolute weapon against the hegemony of global capitalism. Ruwet's interviews of the members of a responsible consumers' association suggests that many of them indeed give a political meaning to their consumption behaviour. Of course, the fact of belonging to an association promoting sustainable consumption is proof that they do not think of their individual resistance as incompatible with collective action: quite the opposite. Besides, as demonstrated by Pirotte, Fair Trade products consumers are members of developmental or environmental associations in greater proportion than are ordinary consumers, and vice versa (80 per cent of volunteers and employees of development cooperation NGOs are consumers of this type of product).

Basically, this consumer/citizen distinction includes a conceptual pair that is much more ancient, the 'private/public' couple, and the question could be asked: Does the responsible consumer that pays up to as much as 10 per cent more for a Fair Trade product (see De Pelsmacker *et al.*, Chapter 8, this volume) act in a purely private way or is it rather public action?

If we adopt Dewey's conception of the public (1927), it seems we must consider that it is a public act. For Dewey, indeed, an action, a behaviour or a transaction ceases to be private as of the moment when they have indirect consequences of certain importance for individuals not taking part in the transaction. In other words, transactions or actions of which the consequences affect groups or individuals other than those who are directly involved fall into the public sphere. If a 'public' is established around these consequences, these then become subject to regulations and control. Conversely, as soon as they no longer bear direct consequences, activities previously considered as dependent of the public domain can return to the private sphere. It is thus, for example, that according to this interpretation, religious rites and beliefs have gone from the public to the private sphere when the members of a social community have ceased to believe that the consequences of one's piety or impiety can affect the community.

It is clear that, through its environmental and social consequences in the context of globalization, an important part of household consumption takes, from now on, an increasingly public character. It is in these terms that the slow but inexorable progress of responsible consumption must be considered as a consumption that has become aware of its henceforth public character.

This is because, beyond the different forms that this responsible consumption can take in terms of ecological or Fair Trade products, of voluntary simplicity or ethical investments, what these 'consumer actors' ask for is, precisely, awareness. The responsible consumers interviewed by Ruwet define themselves as 'autonomous, free, aware' as opposed to the modal consumer perceived as 'dependent, compulsive, unaware'.

Awareness and morality

Responsible consumption cannot be reduced to the affirmation of liberty and reflexivity. It also includes a moral dimension and this is how it can be differentiated not only from the rational consumer as theorized by the neo-classical economy, and who we can believe exists solely in micro-economic books, but it can also be differentiated from a consumption figure that is as real as it gets: that of the *sensible* consumer. Indeed between the rational calculator shaped by the neoclassical economy that chooses in all freedom (without the least external influence) the basket of consumption goods that will provide him or her with as optimal a utility as possible when taking into account budgetary constraints, and the ultra-socialized creature, alienated by publicity, bent 'Lashed by Pleasure, this merciless murderer' (Baudelaire) of the despisers of consumption society, exists and has always existed, a majority of consumers are responsible, not in the meaning given here but, let us say, sensible. The sensible consumer is the one who is approached by consumer associations and that these associations, at the same time, contribute to bring about. Besides, upon closer look, thoughtless and compulsive consumption has never been suggested as an example to follow. It is, by the way, assimilated to adolescence and youth, to the spoiled child and, by this, to immaturity and irresponsibility inherent to this transitory period of life. The consumerist ideal transmitted by the family, the school, the state, the media (think only of the popular female magazines), by trade unions, female associations and consumer associations, is not and has never been the compulsive consumer, the fashion victim, but contrarily, the sensible housewife who manages the family budget with moderation and a sense of thrift, who seizes the opportunities offerered via promotions, discounts, oddments, and worries first about the health and well-being of her family. To a certain extent, *that* consumer has always been fundamentally responsible. Is this to say he or she is moral?

In a way, Miller is right to state that 'at present, consumption is, contrary to most assumptions, a highly moral activity. But this is a morality based largely around the ethical issues of home and family' (1995: 48). This sensible con-sumer's moral is limited to the private sphere. It is that of a consumer who has not yet measured the increasingly public character of consumption.

Responsible or sustainable?

The responsible consumer is thus the one who, having become aware of the public character of his or her consumption and of its impacts on others (directly or indirectly), subordinates his or her consumption choices to considerations other than the simple satisfaction of his or her needs and desires. These considerations obviously have a fundamentally ethical character. In his analysis of the motivations of consumers of Fair Trade products, Pirotte establishes a distinction between three ethical categories that can each be the basis of responsible consumption: assistance, solidarity and justice. Assistance, or help, is defined as a unilateral relationship based on the recognition of the suffering of the other, without the notion (or the possibility) of reciprocity, contrary to solidarity, which is established between equals and implies therefore the possibility of reciprocity. Finally, justice would be based on the idea of inalienable rights and universal standards. Pirotte then notes that the regular customers of Fair Trade stores fit more into a solidarity and justice perspective than the average consumer who rather adheres to a conception of ethics based on the notion of help and assistance.

This typology evokes that of Kohlberg (1981). We know the latter worked out a typology (and a hierarchy) of the stages of moral development in the child in six stages characterized by the passage of anomia (pre-conventional moral) to heteronomy (conventional moral), then to autonomy (post-conventional moral). It is obviously not possible to start here too exhaustive a discussion on the validity and universality of the model of Kohlberg. We know that it does not necessarily provoke the unanimity, even if authors as prestigious as Habermas, for example, refer to it on many occasions. Simply, it seems to us, at least, to constitute a convenient tool to classify the subjacent motivations to consumption. It is what we have tried to do in Table 16.1.

We can see that the motivations of assistance, solidarity and justice evoked by Pirotte coincide rather well with stages 5 and 6 of Kohlberg. On their side, Rousseau and Bontinckx come up with a certain number of motivations which could also find their place in the typology of Kohlberg: comfort, protection and health correspond rather well to Stage 2, the feeling of group membership to Stage 3, normality and 'good citizenship' to Stage 4, altruism to Stage 5. Ruwet's tradimodern consumer would be at Stage 2, the committed and supportive type at Stage 5 and the 'inspired' type at Stage 6.

On the other hand, Dobré's ordinary resistant – who Ruwet however seems to want to bring closer to her inspired type of consumer – slips with difficulty into the grid: concerned only with itself, it seeks neither solidarity nor universal justice but hopes to auto-determine itself, or at least to resist the pressures of the system, be it through non-consumption.

Yet, if responsibility is truly a sufficient condition of durability, it does not constitute a necessary condition of it. Indeed, what determines the durability of consumption is not the intention that guides it but the objective reper- cussions it involves. In other words, the sustainable consumer is the one whose consumption, consciously or not, voluntarily or not, is compatible with the satisfaction of the needs of all the current and future inhabitants of the planet. In this respect, it is compatible with all the levels of moral conscience except the first. Thus one should admit that consumption can be sustainable without being *responsible* in the sense usually given to this term, i.e. without being guided by an ethical intention. The consumption of biological agriculture products, for example, can be motivated exclusively by hygienist and sanitary considerations independently of any environmental concern. In the same way, if the new marketing analysed by Le Velly and used to promote Fair Trade products achieves its goals, it is probable that their consumption will no longer come mainly from responsible consumption.

If, for any reason, one has to expect that only a minority of the population can behave in its consumption in a responsible way (i.e. in terms correspond- ing to Kohlberg's stages 5 and 6), then it is to be feared that sustainable

Table 16.1 Kohlberg's stages of moral development and morality applied to consumption

Moral stage	Description	Moral of consumption
Level 1: Pre-conventional moral (anomia)	1 No rules: obedience to authority, fear of punishment	None: anomic consumption, frantic, bulimic (no internal limits)
	2 Search of personal interest (instrumental relativism – rise in well-being and convenience)	Search of personal well-being: egoistic consumption, utilitarianism (*homo oeconomicus*)
Level 2: Conventional moral (heteronomy)	3 Search for group approval, for agreement with others ('Satisfy expectations of the social environment')	Consumption as a display of group membership, or cultural belonging
	4 Respect of laws, sense of duty ('Respect of social rules')	Regulated and proper consumption (no excess, no waste, reasonable) ('Good housewife', 'Good head of family')
Level 3: Post-conventional moral (autonomy)	5 Care for others' well-being ('Search for the greater good for the most people')	Solidary consumption: Fair Trade and so on
	6 Reflection about ethical principles and search for justice (categorical imperative)	Responsible consumption as consumption submitted to 'intentional duty' (Jonas)

Source: Applied and adapted from Kohlberg (1981).

development remains forever out of reach. Indeed, the latter requires that the models of sustainable consumption become the standard and not the exception, a fact of the mass and not one of a conscious minority. If this is the case, it is clear that something will have to be done so that sustainable consumption (in terms of repercussions) becomes compatible with an ethics of a morally less demanding consumption, compatible even with Stage 2: the research of the well-understood personal interest. Besides, it is what communication seeks to realize in terms of energy consumption, for example, by trying to persuade the consumer that to save energy is 'good for his or her wallet' while preserving his or her comfort and well-being and so on. To a certain extent, as Le Velly has shown, it also appears to be in this direction that marketing for Fair Trade products seems to evolve.

However, another strategy is possible. Undoubtedly more difficult, probably longer, but perhaps no more hazardous, it is that which would seek to make the greater number of consumers responsible in the full meaning of the word.

Criteria of ecological and social sustainability

If important problems arise, as we can see, to an extension of more responsible consumption modes in their intentional aspect, their sustainable slope with the objective meaning as defined above also reveals multiple difficulties. Indeed, if one wants to be rigorous, it appears impossible to define undeniable criteria available on a case-by-case basis of what could be such consumption.

Let us first think of the ecological criteria. That the withdrawal of natural resources be no higher than the possibilities of reconstitution, which is one of the basic rules of ecological sustainability, so be it. But how is one to measure it for a given consumption: this fruit, this clothing, that distance travelled by car? One can eventually do this only with the help of complex methods, partly abstract, of estimates of the repercussions, such as the life-cycle analysis, the ecological footprint, the total material requirement, and especially while multiplying such-and-such consumption by a factor that would allow its hypothetical generalization in the world, another rule sometimes used to estimate sustainability (Sachs *et al.*, 1998). Boulanger (Chapter 2, this volume) resorts to another method of aggregation via the national 'genuine savings'; however, that does not modify the problem of fixing criteria concerning a given consumption.

The truth is rather that many actual systems of production and consumption are, through some aspects, untenable ecologically, and that forecasts allow an aggravation of these impacts for certain essential aspects of the environment to come to light, though certain problems vary strongly from one place to another. Thus sustainable consumption in an Indian province, which suffers an increase in water shortage, first implicates a parsimonious use of water. One could multiply such examples at the local level.

However, this is not how the problem of sustainable consumption found its principal terms of reference. It emerged from societies with an abundance of goods, where the products for human consumption are largely disconnected from the local environmental limits, which makes us seek general criteria. These alternative standards are in conformity with the logic of a potentially globalized market. The militant consumer enjoys a purchasing power that allows him or her a certain number of choices, which brings hope that he can not only voluntarily limit the repercussions, but even influence the organization of the market itself. This remains very partial and limited at the present time.

The call for production, for exchange and consumption in a political influence objective could exist outside societies with an abundance of goods (e.g. Quakers, boycotts launched by Gandhi in India), but in a restricted or temporary way. The call for sustainable consumption represents an ambitious attempt, based on consumers, to solve or at least to look after the vast problem of non-sustainability of development. We evoked the ambiguity of ecological criteria adopted on a case-by-case basis. The situation is even fuzzier in the field of socioeconomic criteria of production and exchange. Nowadays, many developing countries expand production capacities, enabling them to compete with the great powers of this world by asserting fewer obstacles with the exploitation of their comparative advantages, those of a labour where the wages are in phase with the economic level of these countries, and sometimes with working conditions reaching the same desperate level as those same great powers at the time when it colonized them. What would be the attitude to be adopted by a sustainable consumer with regard to these aspects that are today crucial political questions in rich countries? Beyond answers in terms of a selective choice of Southern-based production plants, which respect certain criteria on working conditions (when it is possible), would the logic of its action go so far as to help, by its purchases, these dominated countries to grow richer by an increase in their industrial production, be it to the detriment of less qualified employments in the consumer's own area?

Adding to the difficulties of a definition of ecological criteria, we find here an important example (even if it is little conceptualized in the movement of sustainable consumption) of the difficulty of determining just criteria on social matters, although sustainable development claims to seek a greater global justice. We will also notice, by asking ourselves these questions, that the products coming from the current Fair Trade, stemming from agricultural and craft productions, do not enter, as far as they are involved, in competition with productions from industrialized countries, and can even be included in the line of products sold by companies belonging to these countries.

Ecological criteria, social criteria . . . what about economic criteria, if we want to follow the triptych according to which sustainable development is often presented? The current prices constitute concrete regulations by which the problems mentioned above are dealt with in the market system. They are supposed to integrate the costs and benefits at all the stages of the chain. Of course, ecological criticism has, for a long time, pointed to an insufficient internalization of the environmental costs in the fixing of prices. In the same way, an abundant and sharp social critic, within the framework of alter-globalization, has denounced the huge differences between a price that would be fair and the current prices (see also Boulanger, Chapter 2, this volume, for the prices of raw materials). We certainly cannot look further into these questions here, but we should recognize that they are asked with the help of the undertakings of sustainable consumption. These, in their underlying principle, include a questioning of the economic regulations leading to the current prices, but with all the difficulties in setting up the alternatives in a market system.

Reality and the future of sustainable consumption

However, in spite of the weakness of its standards in technical terms, there is a strong normative sense attached to sustainable consumption today. This contradiction may be understood by the fact that its normalcy, or its ideality, aims in turn for the difficulty present in the existential and moral existence of man in a world of merchandise at the same time as the desired correction of vast socioeconomic systems. From here stems the strength of militant calls (Latouche), even if they do not today fit into a sufficiently coherent and organized approach to create hope for better sustainability.

Yet sustainable consumption is not only about ideals and contradictions. It is also a reality made up of approaches which, in the field, test these contradictions, these ideals, and the complexity of men and of societies.

Reality is when it explores the competing standards and ideals that model the present and the future of Fair Trade. Against the difficulties in fixing social criteria, companies, associations and various actors show how, in their interactions between North and South, these standards can be elaborated (Scherer-Haynes, Le Velly). Reality is when professional methods attempt to include consumers in the promotion of more ecological goods (Rousseau and Bontinckx, Bartiaux), or resulting from Fair Trade (De Pelsmacker *et al.*). Reality still when analyses seek coherent criteria that could gradually orient companies (Fraselle and Scherer-Haynes) and policies in this direction (Uiterkamp, Boulanger, Wallenborn).

Can sustainable consumption become a politicized question, we wondered in the introduction to this work? Nowadays with a limited range in public policies (Reisch and Ropke, 2004), it is difficult to foresee how this topic will evolve as a structuring objective. But it is probable that if it gains size, more questions about the accuracy of the transfer of consumption repercussions, as well as the relations between consumption and the 'good life', will be asked. Broad and numerous as they are, much time will pass before they become exhausted.

We wanted to contribute to the advancement of this question, spurred by the various untenable aspects of our societies. At the end of our attempt, if the limits of the role of consumers arise sometimes by way of contrast, it is in no way a condemnation, a broad point of view denying any validity to these endeavours. If some of the contradictions and difficulties indicated were to progress towards more equitable resolutions, which we would wish, it is certainly not through solutions discussed behind closed doors, but mainly through concrete effort, of which the consumers, among other actors, can really take a share. In the end, after our inquiries on consumers' roles, one could perhaps ask the question 'Sustainable consumption: what role for researchers?' And our answer would be, far from exhausting the subject, to reflect these actions, in order to advance the benefit of the fundamental objectives of equitable or sustainable development. It is thus here, in spite of the severe limitations revealed within these movements, a work designed to support the relevant actions in this direction, to which we wanted to contribute.

Bibliography

Ajzen, I. (1985) 'From intentions to actions: a theory of planned behaviour', in J. Kuhl and J. Beckman (eds) *Action Control: From Cognition to Behaviour*, Heidelberg: Springer.

Ajzen, I. and Fishbein, M. (1980) *Understanding Attitudes and Predicting Social Behaviour*, Englewood Cliffs, NJ: Prentice-Hall.

Anders, G. (1956) *Die Antiquiertheit des Menschen*, trans. (2001) *L'obsolescence de l'homme*, Paris: Ivréa.

Appaduraï, A. (1986) 'Introduction: commodities and the politics of value', in A. Appaduraï (ed.) *The Social Life of Things*, Cambridge: Cambridge University Press.

Argenti, P. (2004) 'Collaborating with activists: how Starbucks works with NGO's', *California Management Review*, 47: 91–116.

Argyle, M. (1999) 'Causes and correlates of happiness', in D. Kahneman, E. Diener and N. Schwarz (eds) *Well-being: The Foundation of Hedonic Psychology*, New York: Russell Sage Foundation.

Arrow, K., Daily, G., Dasgupta, P., Ehrlich, P., Goulder, L., Heaf, G., Levin, S., Mäler, K-G., Schneider, S., Starrett, D. and Walker, B. (2002) *Are We Consuming Too Much?*, Stockholm: Beijer International Institute of Ecological Economics.

Ayres, R.U. (1998) *Turning Point: The End of the Growth Paradigm*, London: Earthscan.

Barbour, R.S. and Kitzinger, J. (eds) (1999) *Developing Focus Group Research: Politics Theory and Practice*, London: Sage.

Bartiaux, F. (2002) 'Relégation et identité: les déchets domestiques et la sphère privée', in P. Magali (ed.) *Déchets rejetés, identités aménagées: Approches sociologiques*, Paris: L'Harmattan.

Bartiaux, F. (2003) 'A socio-anthropological approach to energy-related behaviours and innovations at the household level', in S. Attali, E. Métreau, M. Prône and K. Tillerson (eds) *Time to Turn Down Energy Demand, ECEEE Summer Study Proceedings*, Saint-Raphaël, France, June.

Bartiaux, F. (2005) 'Fragmented rationales and euphemistic behaviours in everyday consumption', Paper presented at the International Sociological Association, Research Committee 24, Environment and Society, *Double Standards and Simulation: Symbolism, Rhetoric and Irony in Eco-Politics*, University of Bath, September.

Bartiaux, F. and Gram-Hanssen, K. (2005) 'Socio-political factors influencing household electricity consumption: A comparison between Denmark and Belgium', in S. Attali and K. Tillerson (eds) *Energy Savings: What Works and Who Delivers, ECEEE Summer Study Proceedings*, Mandelieu La Napoule, France, June 2005.

Baudrillard, J. (1970) *La société de consommation*, trans. (1998) *The consumer society*, London: Sage.

Bauman, Z. (2000) *Liquid Modernity*, Cambridge and Oxford: Polity Press.

Beck, U. (1986; 2nd edn 2001) *La société du risque*, trans. (1986, 1992) *Risk Society, Towards a New Modernity*, London: Sage.

Beck, U. (1987) 'The anthropological shock: Chernobyl and the contours of the risk society', *Berkeley Journal of Sociology*, 32: 153–165.

Beck, U. (1992) 'From industrial society to the risk society: questions of survival, social structure and ecological enlightenment', *Theory, Culture and Society*, 9: 97–123.

Beck, U. (2003) *Pouvoir et contre-pouvoir à l'ère de la mondialisation*, Paris: Aubier.

Bernard, M., Cheynet, V. and Clémentin, B. (eds) (2003) *Objectif décroissance. Vers une société harmonieuse*, Paris: Paragon.

Bird, K. and Hughes, D.R. (1997) 'Ethical consumerism: the case of 'Fairly-Traded' coffee', *Business Ethics: A European Review*, 6: 159–167.

Blokhuis, H. J., Jones, R. B., Geers, R., Miele, M. and Veissier, I. (2003) 'Measuring and monitoring animal welfare: transparency in the food product quality chain', *Animal Welfare*, 12: 445–455.

Boardman, B., Palmer, J. and Lane, K. (2003) *4CE: Consumer Choice and Carbon Consciousness: Electricity Disclosure in Europe*, Final project report, Environmental Change Institute, University of Oxford.

Bode, C. (ed.) (1977) *The Portable Thoreau*, London: Penguin.

Boltanski, L. and Thévenot, L. (1991) *De la justification. Les économies de la grandeur*, trans. (2006) *On Justification: Economies of Worth*, Princeton, NJ: Princeton University Press.

Bordwell, M. (2002) 'Jamming culture: adbusters' hip media campaign against consumerism', in T. Princen, M. Maniates and K. Conca (eds) *Confronting Consumption*, Cambridge, MA: MIT Press.

Bourdieu, P. (1979) *La distinction. Critique sociale du jugement*, trans (1984) *Distinction – A Social Critique of the Judgement of Taste*, London: Routledge.

Bourdieu, P. (1992) *Les Règles de l'art*, Paris: Seuil.

Bourdieu, P. (1997) *Méditations pascaliennes*, Paris: Seuil.

Bowen, H.R. (1953) *Social Responsibilities of the Businessman*, New York: Harper & Row.

Boy, D. (1999) *Le progrès en procès*, Paris: Presses de la Rennaissance.

Bozonnet, J-P. (2005) 'L'écologisme autrement? Fin du grand récit et désinstitutionalisation', Paper presented at ACFAS 73rd international conference, *Nouveaux mouvements sociaux économiques et développement durable: les nouvelles mobilisations à l'ère de la mondialisation*, Chicoutimi, Canada, May.

Brand, K-W. (1997) 'Environmental consciousness and behaviour: the greening of lifestyles', in M. Redclift and G. Woodgate (eds) *The International Handbook of Environmental Sociology*, London: Edward Elgar.

Brewer, J. and Porter, R. (eds) (1993) *Consumption and the World of Goods*, London: Routledge.

Brown, P. and Cameron, L. (2000) 'What can be done to reduce overconsumption?', *Ecological Economics*, 32: 27–41.

Brune, F. (2004) *De l'idéologie aujourd'hui*, Lyon: Parangon.

Bruyer, V., Delabaere, P., Kestemont, M-P., Rousseau, C., Wallenborn, G. and Zaccaï, E. (2004) *Critères et Impulsions de changements vers une consommation durable: approche sectorielle*, Brussels: Belgian Science Policy.

Burgess, J. (2003) 'Sustainable consumption: is it really achievable?', *Consumer Policy Review*, 11: 78–84.

Byrd-Bredbenner, C. and Coltee, A.W.A.P. (2000) 'Consumer understanding of US and EU nutrition labels', *British Food Journal*, 102: 615–629.

Cabin, P., Desjeux, D., Nourisson, D. and Rochefort, R. (1998) *Comprendre le consommateur*, Auxerre: Sciences Humaines.

Callon, M. (1995) 'Four models for the dynamics of science', in S. Jasanoff, G. Markle, J. Petersen and T. Pinch (eds) *Handbook of Science and Technology Studies*, London: Sage.

Callon, M. (1998) 'Introduction: the embeddedness of economic markets in economics', in M. Callon (ed.) *The Laws of the Markets*, Oxford: Blackwell.

Campbell, C. (1990) *The Romantic Ethic and the Spirit of Modern Consumerism*, London: Blackwell.

Campbell, C. (1995) 'The sociology of consumption', in D. Miller (ed.) *Acknowledging Consumption*, London: Routledge.

Caradec, V. and Martuccelli, D. (2004) *Matériaux pour une sociologie de l'individu*, Villeneuve d'Ascq: Presses Universitaires du Septentrion.

Carr-Hill, R.A. and Lintott, J. (2002) *Consumption, Jobs and the Environment: A Fourth Way?*, London: Palgrave Macmillan.

Carrigan, M. and Attalla, A. (2001) 'The myth of the ethical consumer – do ethics matter in purchase behaviour?', *Journal of Consumer Marketing*, 18: 560–577.

Caswell, J.A. and Mojduszka, E.M. (1996) 'Using informational labelling to influence the market for quality in food products', *American Journal of Agricultural Economics*, 78: 1248–1253.

Caswell, J.A. and Padberg, D.I. (1992) 'Toward a more comprehensive theory of food labels', *American Journal of Agricultural Economics*, 74: 460–468.

Charlier, S., Yépez del Castillo, I. and Andia, E. (2000) *Payer un juste prix aux cultivatrices de quinoa*, Brussels: Luc Pire.

Chessel, M.E. and Cochoy, F. (2004) 'Autour de la consommation engagée. Enjeux historiques et politique', *Sciences de la Société*, 64: 3–14.

Chichilnisky, G. (1994) 'North–South trade and the global environment', *American Economic Review*, 84: 851–874.

Chosson, A. (2002) 'Consommateur-citoyen ou "usager" du développement durable?', *Autrement*, 216: 163–168.

CNUCED (2005) *Informations sur le coton*. UNCTAD. Online. Available: http://r0. unctad.org/infocomm/francais/coton/plan.htm (accessed 23 April 2006).

Cogoy, M. (1999) 'The consumer as a social and environmental actor', *Ecological Economics*, 28: 385–398.

Cohen, M. and Murphy, J. (eds) (2001) *Exploring Sustainable Consumption. Environmental Policy and the Social Sciences*, Amsterdam and London: Pergamon Press.

Cole, H.S.D., Freeman, C., Jahoda, M. and Pavitt, K.L.R. (1973) *Thinking About the Future: A Critique of the Limits to Growth*, London: Chatto & Windus.

Collomb, P. and Guérin-Pace, F. (1998) *Les Français et l'environnement. L'enquête 'Populations-Espaces de vie-Environnements'*, Paris: Presses Universitaires de France.

'Consumption, Everyday Life and Sustainability program', European Science Foundation, Lancaster.

Corcuff, P. (1995) *Les nouvelles sociologies: Constructions de la réalité sociale*, Paris: Nathan.

Corcuff, P. (2001) 'Le collectif au defi du singulier: en partant de l'habitus', in B. Lahire (ed.) *Le travail sociologique de Pierre Bordieu: Dettes et critiques*, Paris: La Decouverte/ Poche (Sciences humaines et sociales).

Corcuff, P. (2002) 'Respect critique', in *L'oeuvre de Pierre Bourdieu*, Auxerre: Sciences Humaines.

Coriat, B. (1988) *L'atelier et le chronomètre*, Paris: Christian Bourgois.

CRC-Consommation (1998) *Commerce ethique: les consommateurs solidaires*, Paris: CRC.

CRIOC (2004) Perception des labels, Brussels: CRIOC. Online. Available: http:// www.crioc.be (accessed November 2005).

Crocker, D.A. and Linden, T. (1998) 'Introduction', in D.A. Crocker and T. Linden (eds) *Ethics of Consumption. The Good Life, Justice and Global Stewardship*, Lanham, MD: Rowman & Littlefield.

Cross, G. (1993) *Time and Money: The Making of Consumer Culture*, London: Routledge.

Cross, J.G. and Guyer, M.J. (1980) *Social Traps*, Ann Arbor, MI: University of Michigan Press.

Crozier, M. and Friedberg, E. (1977) *L'acteur et le système*, Paris: Points.

Dalla Valle, C., MONS, J., Bartiaux, F. and Yzerbyt, V. (2001) 'Les actions environ-nementales des consommateurs', in M-P. Kestemont, F. Bartiaux, N. Fraselle and V. Yzerbyt (eds) *Points d'ancrage pour une politique de développement durable: production et consommation*, Brussels: Belgian Science Policy.

Daly, H.E. (1996a) *Beyond Growth: The Economics of Sustainable Development*, Boston, MA: Beacon Press.

Daly, H.E. (1996b) 'Physical growth versus technological development', in D.C. Pirages (ed.) *Building Sustainable Societies*, Armonk and London: M. E. Sharpe.

De Cenival, L. and SOLAGRAL (1998) *Du commerce équitable à la consommation responsable*, Maastricht: EFTA.

De Certeau, M. (1980) *L'invention du quotidien, Tome 1 – Arts de faire*, Paris: Gallimard.

De Pelsmacker, P., Driessen, L. and Rayp, G. (2005) 'Do consumers care about ethics? Willingness-to-pay for Fair-Trade coffee', *Journal of Consumer Affairs*, 39: 361–383.

De Pelsmacker, P., Janssens, W., Sterckx, E. and Mielants, C. (2005a) 'Fair Trade beliefs, attitudes and buying behaviour of Belgian consumers', *Journal of Voluntary Sector and Non-profit Marketing*, 22: 512–530.

De Pelsmacker, P., Janssens, W., Sterckx, E. and Mielants, C. (2005b) 'Consumer preferences for different approaches to marketing ethical-labelled coffee', *International Marketing Review*, 22: 512–530.

Deaton, A. (2003) *Regional Poverty Estimates for India, 1999–2000*, Research Program in Development Studies, Princeton University. Online. Available: http://www. wws.princeton.edu/~rpds/downloads/deaton_regionalpovertyindia.pdf (accessed April 2005).

Deleuze, G. (1990) *Pourparlers (1972–1990)*, Paris: Editions de Minuit.

Descola, P. (1999) 'Ecologiques', in P. Descola, J. Hamel and P. Lemonnier (eds) *La production du social. Autour de Maurice Godelie*, Paris: Fayard.

Detienne, M. and Vernant, J-P. (1974) *Les ruses de l'intelligence – La mètis des Grecs*, Paris: Champs Flammarion.

Dewey, J. (1927) *The Public and Its Problems*. Denver, CO: Allan Swallow.

Diamantopoulos, A., Schlegelmilch, B., Sinkovics, R. and Bohlen, G. (2003) 'Can socio-demographics still play a role in profiling green consumers? A review of the evidence and an empirical investigation', *Journal of Business Research*, 56: 465–480.

Dickson, M.A. (2001) 'Utility of no-sweat labels for apparel consumers: profiling label users and predicting their purchases', *Journal of Consumer Affairs*, 35: 96–119.

Diener, E. (1984) 'Subjective well-being', *Psychological Bulletin*, 95: 542–575.

Dobré, M. (2002) *L'écologie au quotidien. Eléments pour une théorie sociologique de la résistance au quotidien*, Paris: L'Harmattan.

Dobré, M. and Lianos, M. (2002) 'Modernité et vulnérabilité: incertitude, insécurité et retrait de la société en France', *Europaea, Journal of the Europeanists*, 1–2: 319–349.

Dolan, P. (2002) 'The sustainability of "sustainable consumption"', *Journal of Macromarketing*, 22: 170–181.

Dominguez, J. and Robin, V. (1992) *Your Money or Your Life*, Harmondsworth: Penguin.

Dortier, J-F. (1998) 'L'individu dispersé et ses identities multiples', in J.-C. Ruano-Borbalau (eds) *L'identité, l'individu, le groupe, la societé*, Auxerre: Editions Sciences Humaines.

Douglas, M. (1983) *Risk and Culture: An Essay on the Selection of Technical and Environmental Dangers*, Berkeley, CA: University of California Press.

Douglas, M. (1992) 'In defence of shopping', in R. Eisende and E. Miklautz (eds), *Produktkulturen, Dynamik und Bedeutungswandel des Konsums*, Frankfurt and New York: Kampus.

Douglas, M. (1993) 'A quelles conditions un ascétisme environnementaliste peut-il réussir?', in D. Bourg (ed.) *La nature en politique, ou l'enjeu philosophique de l'écologie*, Paris: L'Harmattan.

Douglas, M. and Isherwood, B. (1979) *The World of Goods: Towards an Anthropology of Consumption*, London: Routledge.

Durning, A. (1993) *How Much is Enough?*, New York: Norton & Co.

Easterlin, R. (1995) 'Will raising the incomes of all increase the happiness of all?', *Journal of Economic Behaviour and Organization*, 27: 35–47.

Easterlin, R. A. (1972) 'Does economic growth improve the human lot? Some empirical evidence', in P.A. David and M.W. Reder (eds) *Nations and Households in Economic Growth*, Stanford, CA: Stanford University Press.

Eberhart, C. and Chaveau, C. (2002) *Etude du commerce équitable dans la filière café en Bolivie*, Paris: Centre International de Coopération pour le Développement Agricole.

Ehrlich, P. and Holdren, J. (1971) 'Impact of population growth', *Science*, 171: 1212–1217.

Elster, J. (1985) *Making Sense of Marx*, Cambridge: Cambridge University Press.

EC (European Commission) (2004) *Attitudes des citoyens européens vis-à-vis de l'environnement*, Eurobaromètre spécial 217, April 2005.

EEA (European Environment Agency) (2005) *Household Consumption and the Environment*, Technical report N11, Copenhagen: EEA.

EFTA (European Fair Trade Association) (1998) *Fair Trade in Europe. Facts and Figures on the Fair Trade Sector in 16 European Countries*, Maastricht: EFTA.

EFTA (European Fair Trade Association) (2000) *Commerce équitable, Mémento pour l'an 2000*, Maastricht: EFTA.

EFTA (European Fair Trade Association) (2001) *Fair Trade Definition*, Maastricht: EFTA. Online. Available: http://www.eftafairtrade.org/definition.asp (accessed January 2006).

Everett, J. (2001) *Ethics of Consumption: Individual Responsibilities in a Consumer Society*, unpublished thesis, University of Colorado, Boulder.

Fair Trade Labelling Organisation (2004) *Fees for Initial Fair Trade Certification*, Online. Available: http://www.fairtrade.net/index.html (accessed June 2004).

Fair Trade Labelling Organisation (2005) *Fair Trade Standards for Contract Production Projects (India and Pakistan)*, Online. Available: http://www.fairtrade.net/pdf/sp/english/Contract%20Production%20standards.pdf (accessed 4 December 2005).

Fair Trade Labelling Organisation (2006) *Fair Trade Standards for SEED Cotton (Small Farmers)*, Online. Available: http://www.fairtrade.net/pdf/sp/english/Coffee%20SP%20versionJune04.pdf (accessed 10 April 2006).

Flavin, C. and Lenssen, N. (1995) *Power Surge*, London: Earthscan.

Fligstein, N. (2001) *The Architecture of Markets: An Economic Sociology of Twenty-first Century Capitalist Societies*, Princeton, NJ: Princeton University Press.

Foucault, M. (1982) 'Le sujet et le pouvoir', reprinted in *Dits et Ecrits*, 4, 1980–1988 (1994), Paris: Gallimard.

Frank, R.H. (1997) 'The frame of reference as a public good', *Economic Journal*, 107: 1832–1847.

Frank, R.H. (1999) *Luxury Fever: Money and Happiness in an Era of Excess*, Princeton, NJ, and Oxford: Princeton University Press.

Fraselle, N. (2003) *Pour une conscience globale de la consommation*, Brussels: Editions Labor.

Frey, B.S. and Stutzer, A. (2002) 'What can economists learn from happiness research?', *Journal of Economic Literature*, 40: 402–435.

Fuchs, D. and Lorek, S. (2001) *An Inquiry into the Impact of Globalization on the Potential for Sustainable Consumption in Households*, Oslo: Prosus.

Fukuyama, F. (1992) *The End of History and the Last Man*, New York: Macmillan.

Gatersleben, B., Steg, L. and Vlek, C. (2002) 'The measurement and determinant of environmentally significant consumer behaviour', *Environment and Behaviour*, 34: 335–362.

Gauthy-Sinechal, M. and Vandercammen, M. (2005) *Etudes de marchés, méthods et outils*, Brussels: de Boeck.

Geels, F. and Kemp, R. (2000) *Transities vanuit sociotechnisch perspectief*, unpublished report, University of Maastricht (MERIT).

Geller, E.S. (2002) 'The challenge of increasing pro-environmental behaviour', in R.B. Bechtel and A. Churchman (eds) *Handbook of Environmental Psychology*, New York: Wiley.

Gendron, C., Turcotte, M-F., Audet, R., de Bellefeuille, S., Lafrance, M-A. and Maurais, J. (2003) 'La consommation comme mobilisation sociale: l'impact des nouveaux mouvements sociaux économiques dans la structure normative des industries', Chaire Économie et Humanisme, Université du Québec à Montréal.

Georg, S. (1999) 'The social shaping of household consumption', *Ecological Economics*, 28: 455–466.

Gesualdi, F. (1999) *Manuale per un consume responsabile. Dal boicottaggio al commercio equo e solidale*, Milan: Feltrinelli.

Giddens, A. (1984) *The Constitution of Society*, Oxford: Polity Press.

Giddens, A. (1994) *Les conséquences de la modernité*, Paris: L'Harmattan.

Gintis, H. (2000) 'Beyond homo economicus: evidence from experimental economics', *Ecological Economics*, 35: 311–322.

Glasbergen, P. (2002) 'Beleidswetenschappelijk milieuonderzoek: profiel, kernvragen en betekenis', *Milieu*, 17: 250–269.

Goldblatt, D. (2003) 'A dynamic structuration approach to information for end-user energy conservation', in S. Attali, E. Métreau, M. Prône and K. Tillerson (eds) *Time* to *Turn Down Energy Demand*, *ECEEE Summer Study Proceedings*, Saint-Raphaël, France, June 2003.

Goodman, M. (2004) 'Reading Fair Trade: political ecology, imaginary and the moral economy of Fair Trade foods', *Political Geography*, 23: 891–915.

Goodman, M. (2005) 'Going on the "charm offensive": Fair Trade consumption and the politics of intention', Paper presented at the Sustainable Consumption Cluster Meeting on Fair Trade, Liège, October.

Gorz, A. (1977) *Ecologie et liberté*, Paris: Galilée.

Gorz, A. (1989) *Critique of Economic Reason*, London: Verso.

Gram-Hanssen, K. (2005) 'Teenage consumption of information and communication technologies', in S. Attali and K. Tillerson (eds) *Energy Savings: What Works and Who Delivers*, *ECEEE Summer Study Proceedings*, Mandelieu La Napoule, France, June.

Gram-Hanssen, K., Bartiaux, F., Cantaert, M. and Jensen, O.M. (2005) 'Energy expert advices: a comparative study on the Danish energy labelling scheme and the Belgian energy assessment procedure', Paper presented at the 7th Conference of the European Sociological Association, Torun, Poland, September.

Gram-Hanssen, K. Bartiaux, F., Jensen, O.M., Cantaert, M. (forthcoming) 'Do home owners use energy labels? A comparison between Denmark and Belgium', submitted to *Energy Policy*.

Green, K. and Vergragt, P. (2002) 'Towards sustainable households: a methodology for developing sustainable technological and social innovations', *Futures*, 34: 381–400.

Green, P.E., Carroll, J.D. and Carmone, F.J. (1978) 'Some new types of fractional factorial designs for marketing experiments', *Research in Marketing*, 1: 99–122.

Green, P.E., Krieger, A.M. and Wind, Y. (2001) 'Thirty years of conjoint analysis: reflections and prospects', *Interfaces*, 31: 56–73.

Grimes, K. M. and Milgram, L. (eds) (2000) *Artisans and Cooperatives. Developing Alternative Trade for the Global Economy*, Tucson, AZ: The University of Arizona Press.

Grin, J. and Van de Graaf, H. (1999) 'Policy instruments, pratique réfléchie et apprentissage. Implications pour la gouvernabilité à long terme et la démocratie. Gestion négociée des territoires et politiques publiques', *Espaces et sociétés*, 97–98.

Habermas, J. (1981) *Theorie des kommunikativen Handels*, trans. (1987) *Théorie de l'agir communicationnel*, Paris: Fayard.

Habermas, J. (1988) *Raison et légitimité. Problème de légitimité dans le capitalisme avancé*, Paris: Payot.

Haesler, A. (2005) 'Penser l'individu? Sur un nécessaire changement de paradigme', EspaceTemps.net, 23.11.2005, Available: http://www.espacestemps.net/document 1726.html (accessed June 2006).

Hagedoorn, J. W. (1996) 'Happiness and self-deception: an old question examined by a new measure of subjective well-being', *Social Indicators Research*, 38: 138–160.

Halkier, B. (2001) 'Routinisation or reflexivity? Consumers and normative claims for environmental consideration', in J. Gronow and A. Warde (eds) *Ordinary Consumption*, London and New York: Routledge.

Hamilton, K. and Clemens, M. (2000) 'Sustaining economic welfare: estimating changes in per capita wealth', *Policy Research Working Paper*, 2498: World Bank.

Hannigan, J.A. (1995) *Environmental Sociology, A Social Constructionist Perspective*, London and New York: Routledge.

Harrison, R., Newholm, T. and Shaw, D. (2005) *The Ethical Consumer*, London: Sage.

Hawken, P., Lovins, A.B. and Lovins, L. H. (1999) *Natural Capitalism: The Next Industrial Revolution*, London: Earthscan.

Hayden, A. (1999) *Sharing the Work, Sparing the Planet*, London: Zed Books.

Haynes, I. (forthcoming) 'Can we keep alive the roots of Fair Trade? Influence of consumption logics on the FT referential', *International Journal of Agricultural Resources, Governance and Ecology*.

Heap, B. and Kent, J. (eds) (2000) *Towards Sustainable Consumption: a European Perspective*, London: The Royal Society.

Herpin, N. (1995) *Sociologie de la consommation*, Paris: La Découverte.

Hertwich, E. G. (2005) 'Consumption and industrial ecology', *Journal of Industrial Ecology*, 9: 144.

Hines, C. and Ames, A. (2000) *Ethical Consumerism. A Research Study Conducted for the Co-operative Bank*, London: MORI.

Hirsch, F. (1977) *Social Limits to Growth*, London and New York: Routledge.

Hirschman, A. O. (1970) *Exit Voice and Loyalty: Responses to Decline in Firms, Organizations, and States*, Cambridge, MA: Harvard University Press.

Hobson, K. (2002) 'Competing discourses of sustainable consumption: does the rationalisation of lifestyles make sense', *Environmental Politics*, 11: 95–120.

Hobson, K. (2003) 'Thinking habits into action: the role of knowledge and process in questioning household consumption practices', *Local Environment*, 8: 95–112.

Hopkins, R. (2000) *Impact Assessment Study of Oxfam Fair Trade*, Oxford: Oxfam Fair Trade Programme.

Hudson, I. and Hudson, M. (2003) 'Removing the veil? Commodity fetishism, fair trade and the environment', *Organization and Environment*, 16: 413–430.

Hughes, A. (2000) 'Retailers, knowledge and changing commodity networks: the case of the cut flower trade', *Geoforum*, 31: 175–190.

Hughes, T. P. (1987) 'The evolution of large technical system', in T. Bijker, T. Hughes and T. Pinch (eds) *The Social Construction of Technological Systems*, Cambridge, MA: MIT Press.

IDEA Consult (2002) *Effectenstudie en opiniepeiling over eerlijke handel in België*, Brussels: Ministry of Foreign Affairs and International Cooperation of Belgium.

Illich, I. (1973) *Tools for Conviviality*, New York: Harper & Row.

Illich, I. (1974) *Energy and Equity*, London: Calder & Boyars.

Ilomen, K. (2001) 'Sociology, consumption and routine', in J. Gronow and A. Warde (eds) *Ordinary Consumption*, London and New York: Routledge.

Inglehart, R. and Klingemann, H-D. (2000) 'Genes, culture, democracy, and happiness', in E. Diener and E.M. Suh (eds) *Culture and Subjective Well-being*, Cambridge, MA: MIT Press.

IFAT (International Fair Trade Association) (2005) *What is Fair Trade?* Online. Available: http://www.ifat.org/whatisft.html (accessed 7 May 2005).

Iversen, T. (1996) *Miljøproblematikken I Hverdaglivet*, Copenhagen: Institute for Anthropology.

Izraelewicz, E. (2005) *Quand la Chine change le monde*, Paris: Grasset.

Jackson, T. (1996) *Material Concerns*, London and New York: Routledge.

Jackson, T. (2006) *The Earthscan Reader on Sustainable Consumption*, London: Earthscan.

Jackson, T. and Michaelis, L. (2003) *Policies for Sustainable Consumption*, Report to the Sustainable Development Commissions, University of Surrey and University of Oxford.

Jacobs, J. (1961) *The Life and Death of Great American Cities*, New York: Vintage Books.

Jacquemain, M. (2002) *La raison névrotique. Individualisme et société*, Brussels: Labor.

Jaffrelot, C. (2005) *Inde: la démocratie par la caste. Histoire d'une mutation socio-politique. 1885–2005*, Paris: Fayard.

Jensen, O.M. (1996) *Affald I boligområder: Livsstil og affaldsvaner*, SBI-Rapport 261, Hørsholm, Denmark.

Jonas, H. (1979, 2nd edn 1990) *Le principe responsabilité. Une éthique pour une civilisation technologique*, Paris: Cerf.

Jonas, H. (1999) *La créativité de l'agir*, Paris: Cerf.

Jones, E.E. and Harris, V.A. (1967) 'The attribution of attitudes', *Journal of Experimental Social Psychology*, 3: 1–24.

Jullien, F. (1996) *Traité de l'efficacité*, Paris: Grasset.

Kahneman, D. (1999) 'Objective happiness', in D. Kahneman, E. Diener and E. Schwartz (eds) *Well-being, The Foundation of Hedonic Psychology*, New York: Russell Sage.

Kant, E. (1788) *Critique of Practical Reason*, trans. T.K. Abbott. Available: http://www. philosophy.eserver.org/kant/critique-of-practical-reason.txt.

Kasemir, B., Dahinden, U., Gerger, Å., Schüle, R., Tabara, D. and Jaeger, C.C. (2000) 'Citizens' perspectives on climate change and energy use', *Global Environmental Change*, 10: 169–184.

Kaufmann, J-C. (1992) *La trame conjugale: analyse du couple par son linge*, Paris: Nathan.

Kaufmann, J-C. (1993) *Sociologie du couple*, Paris: PUF.

Kaufmann, J-C. (1996) *L'entretien compréhensif*, Paris: Nathan.

Kaufmann, J-C. (1997) *Le cœur à l'ouvrage: théorie de l'action ménagère*, Paris: Nathan.

Kaufmann, J-C. (2001) *Ego. Pour une sociologie de l'individu*, Paris: Collin.

Kaufmann, J-C. (2004) *L'invention de soi. Une théorie de l'identité*, Paris: Collin.

Kersting, W. (1998) 'Internationale solidarität', in K. Bayertz (ed.) *Solidarität. Begriff and Problem*, Frankfurt/Main: Suhrkamp.

Kestemont, M-P. (1999) 'European Business Environmental Barometer', Report SUB/97/5005441, Université Catholique de Louvain.

Kestemont, M-P., Bartiaux, F., Fraselle, N. and Yzerbyt, V. (eds) (2001) *Points d'ancrage pour une politique de développement durable: production et consommation*, Brussels: Belgian Science Policy.

Kohlberg, L. (1981) *Essays on Moral Development*, New York, and San Fransisco, CA: Harper & Row.

Kok, R., Falkena, H.J., Benders, R.M.J., Moll, H.C. and Noorman, K.J. (2003) *Household Metabolism in European Countries and Cities*, Groningen: IVEM Research Report.

Kong, N., Salzmann, O., Steger, U. and Ionescu-Somers, A. (2002) 'Moving business/ industry towards sustainable consumption: the role of NGOs', *European Management Journal*, 20: 109–127.

Krier, J.M. (2001) *Fair Trade in Europe 2001: Facts and Figures on the Fair Trade Sector in 18 European Countries*, Research report for EFTA. Geneva, Brussels, Luxembourg: EFTA.

Krier, J.M. (2005) *Fair Trade in Europe 2005. Facts and Figures on Fair Trade in 25 European Countries*, Brussels: FLO, IFTA, NEWS and EFTA.

Lafaye, C. and Thévenot, L. (1993) 'Une justification écologique? Conflits dans l'aménagement de la nature', *Revue française de sociologie*, 24: 495–524.

Lane, R. (2002) *The Loss of Happiness in Market Democracies*, New Haven, CT, and London: Yale University Press.

Latouche, S. (2003) *Justice sans limites*, Paris: Fayard.

Latouche, S. (2004) *Survivre au développement: De la décolonisation de l'imaginaire économique à la construction d'une société alternative*, Paris: Mille et une Nuits.

Latour, B. (1989) *La science en action*, Paris: La Découverte.

Latour, B. (2005) *Rassembling the Social. An Introduction to Actor-network Theory*, Oxford: Oxford University Press.

Layard, R. (2003) *Happiness: Has Social Science a Clue?*, London School of Economics: Lionel Robbins Memorial Lectures.

Layard, R. (2005) *Happiness. Lessons from a New Science*, London: Penguin.

Le Velly, R. (2004) 'Sociologie du marché. Le commerce équitable: des échanges marchands contre le marché et dans le marché', unpublished thesis, Université de Nantes, France.

Lecomte, T. (2003) *Le pari du commerce équitable*, Paris: Editions d'Organisation.

Lhuilier, D. and Cochin, Y. (1999) *Des déchets et des hommes*, Paris: Desclée de Brouwer.

Lianos, M. (2002) *Le nouveau contrôle social*, Paris: L'Harmattan.

Lianos, M. (2005) 'Deleuze, Foucault et le nouveau contrôle social', *Mana*, 12–13: 11–38.

Linderhof, V.G.M. and Kooreman, P. (1998) 'Economic aspects of household metabolism', in K.J. Noorman and T.S. Uiterkamp (eds) *Green Households? Domestic Consumers, Environment and Sustainability*, London: Earthscan.

Lintott, J. (1998) 'Beyond the economics of more: the place of consumption', *Ecological Economics*, 25: 239–248.

Lintott, J. (2004) 'Work in a growing and in a steady state economy', *International Journal of Environment, Workplace, and Employment*, 1: 40–52.

Lintott, J. (2005) 'Evaluating the "Hirsch hypothesis": a comment', *Ecological Economics*, 52: 1–3.

Lipovetsky, G. (2003) 'La société d'hyperconsommation', *le Débat*, 124: 74–98.

Littrell, M.A. and Dickson, M.A. (1999) *Social Responsibility in the Global Market. Fair Trade of Cultural Products*, London: Sage.

Liu, J., Wang, R. and Yang, J. (2005) 'Metabolism and driving forces of Chinese urban household consumption', *Population and Environment*, 26: 325–341.

Livesey, S. and Kearins, K. (2002) 'Transparent and caring corporations?', *Organization and Environment*, 15: 233–258.

Locke, J. (1966) *The Second Treatise On Government*, Oxford: Blackwell.

Loureiro, L.M., McCluskey, J.J. and Mittelhammer, R.C. (2002) 'Will consumers pay a premium for eco-labeled appeals?', *Journal of Consumer Affairs*, 36: 203–219.

Luhmann, N. (1993) *Risk: A Sociological Theory*, New York: Walter de Gruyter.

Lutzenhiser, L. (1993) 'Social and behavioral aspects of energy use', *Annual Review of Energy and Environment*, 18: 247–289.

MacGillivray, A. (2000) *The Fair Share: The Growing Market Share of Green and Ethical Products*, London: New Economics Foundation.

Macnaghten, P. and Urry, J. (1998) *Contested Natures*, London: Sage.

Madelain, C. (2003) 'Brouillons pour l'avenir', *Les Nouveaux Cahiers de l'IUED*, No. 14, Paris and Geneva: PUF.

MAFF (2000) *Consumers Want Action on Food Labels*, Online. Available: http://www.defra.gov.uk/newsrel (accessed 2 February 2005).

Maietta, O.W. (2003) 'The hedonic price of Fair-Trade coffee for the Italian consumer', Paper presented at the International Conference on Agricultural Policy Reform and the WTO, *Where are we Heading?*, Capri, Italy.

Maignan, I. and Ferrell, O.C. (2001) 'Corporate citizenship as a marketing instrument – concepts, evidence and research directions', *European Journal of Marketing*, 35: 457–484.

Maignan, I. and Ferrell, O.C. (2004) 'Corporate social responsibility and marketing: an integrative framework', *Journal of the Academy of Marketing Science*, 32: 3–19.

Manno, J. (2002) 'Commoditization: consumption efficiency and an economy of care and connection', in T. Princen, M. Maniates and K. Conca (eds) *Confronting Consumption*, Cambridge, MA: MIT Press.

Marcuse, H. (1964) *One Dimensional Man*, Boston, MA: Beacon Press.

Mardsen, T. (2000) 'Food matter and the matter of food: towards a new food governance?', *Sociologia Ruralis*, 40: 20–29.

Martinez-Alier, J. (2002) *The Environmentalism of the Poor*, Cheltenham: Edward Elgar.

Martuccelli, D. (2002) *Grammaires de l'individu*, Paris: Gallimard.

Marx, K. (1844) 'Economic and philosophical manuscripts', in K. Marx and F. Engels *Collected Works*, London: Lawrence & Wishart.

Marx, K. (1867, 1909) *Capital. A Critique of Political Economy*, Vol. 1, Chicago, IL: Charles H. Kerr.

Max Havelaar (2004) 'Company information'. Available: http://www.maxhavelaar.be, accessed December 2004.

Max Havelaar France (January 2005) *Le coton équitable première filière non alimentaire de Max Havelaar*, Dossier de presse. Online. Available: http://www.maxhavelaar france.org/produits/DPCoton.pdf (accessed 10 February 2005).

Max-Neef, M. (1995) 'Economic growth and quality of life, a threshold hypothesis', *Ecological Economics*, 15: 115–118.

McGuire, W.J. (1999) *Constructing Social Psychology. Creative and Critical Processes*, Cambridge: Cambridge University Press.

Meadows, D.H., Meadows, D.L., Randers, J. and Behrens, W.W. (1972) *The Limits to Growth*, New York: Universe Books.

Mielants, C., De Pelsmacker, P. and Janssens, W. (2003) *Kennis, houding en gedrag van de Belgen t.a.v. fair trade producten. Conclusies uit vier focusgroepsgesprekken*, unpublished working paper, University of Antwerp, Belgium.

Miles, S. (1998) *Consumerism as a Way of Life*, London: Sage.

Millennium Ecosystem Assessment (MEA) (2005) *Ecosystems and Human Well-being*, Synthesis available at http://www.maweb.org/, (accessed 11 August 2006).

Miller, D. (1995) 'Consumption as the vanguard of history', in D. Miller (ed.) *Acknowledging Consumption. A Review of New Studies*, London and New York: Routledge.

Minsberg, H. (1986) *Le pouvoir dans les organisations*, Paris: Organisations.

Moisander, J. (1999) 'Motivation for ecologically oriented consumer behaviour', Working paper, Second Europa, Science Foundation Workshop on Consumption, Everyday Life and Sustainability, Lancaster University.

MORI (2000) *European Attitudes Towards Corporate Social Responsibility*, CSR Europe, London: MORI.

Mrill, I. (2000) 'Producing the natural fiber naturally: technological change and the US organic cotton industry', *Agriculture and human values*, 17: 325–336.

Murray, D.L., Raynolds, L.T. and Taylor, P.L. (2003) *One Cup at a Time: Poverty, Alleviation and Fair Trade Coffee in Latin America*, Fair Trade Research Group Report, State University of Colorado.

Myers, D.G. (1992) *The Pursuit of Happiness*, New York: Avon Books.

Myers, N. and Kent, J. (eds) (2004) *The New Consumers. The Influence of Affluence on the Environment*, Washington, DC, Covelo, and London: Island Press.

Naess, S. (1994) 'Does self-deception enhance the quality of life?', in L. Nordenfelt (ed.) *Concepts and Measurement of Quality of Life in Health Care*, Dordrecht: Kluwer.

Narayan, J. (1946) *Towards Struggle. Selected Manifestos, Speeches and Writings*, Bombay: Padma Publications.

Nielsen, A.C. (2002) 'Data on the Belgian coffee market (1999–2001)', unpublished technical report and data, Brussels: A.C. Nielsen.

Niestroy, I. (2005) *Sustaining Sustainability*, Background Study No. 2, The Hague: EEAC.

Nilsson, H., Tunçer, B. and Thidell, A. (2004) 'The use of eco-labeling like initiatives on food products to promote quality insurance – is there enough credibility?', *Journal of Cleaner Production*, 12: 517–526.

Nilsson, O.S., Nissen, N.P., Thorgersen, J. and Vilby, K. (1999) 'Rapport fra "forbrugergruppen" under Erhvervsministeriets maerkningsudvalg' (A report from the consumer study group to the Labels Committee), in *Maerkningsudvalgets redegorelse*, Copenhagen: Ministry of Trade and Industry, the Consumer Agency.

Niva, M. and Timonen, P. (2001) 'The role of consumers in product-oriented environmental policy: can the consumer be the driving force for environmental improvements?', *International Journal of Consumer Studies*, 25: 331–338.

Noorman, K.J. and Schoot Uiterkamp, T. (eds) (1998) *Green Households? Domestic Consumers, Environment and Sustainability*, London: Earthscan.

Norton, B., Costanza, R. and Bishop, R. (1998) 'The evolution of preferences. Why "sovereign" preferences may not lead to sustainable policies and what to do about it', *Ecological Economics*, 24: 193–211.

Nozick, R. (1974) *Anarchy, State and Utopia*, Cambridge, MA: Harvard University Press.

OECD (2002a) *Towards Sustainable Household Consumption? Trends and Policies in OECD Countries*, Paris: OECD.

OECD (2002b) *Indicators to Measure Decoupling of Environmental Pressure from Economic Growth*, Paris: OECD.

Oosterhuis, F., Rubik, F. and Scholl, G. (1996) *Product Policy in Europe: New Environmental Perspectives*, Dordrecht, Boston: MA, and London: Kluwer.

Oswald, A.J. (1997) 'Happiness and economic performance', *Economic Journal*, 107: 1815–1831.

Otnes, P. (1988) *The Sociology of Consumption*, Paris: Humanities press International.

Oxfam (2003) http://www.oxfam.be/, (accessed December 2003).

OPM/SMG (Oxford Policy Management, Sustainable Markets Group) (2000) *Fair Trade: Overview, Impact and Challenge*, Oxford: OPM.

Pacteau, C. (1999) 'Penser. De la logique à l'expérience', in J-F. Dortier (ed.) *Le cerveau et la pensée. La révolution des sciences cognitives*, Auxerre: Sciences Humaines Éditions.

Pallante, M. (2005) *La decrescita felice*, Rome: Editori Riunti.

Pallemaerts, M., McLoughin, A., Jans, M.T., Bedoyan, I., Tanasescu, I., Misonne, D. and Petitat V. (2006) *The Role of Public Authorities in Integrated Product Policy: Regulators or Coordinators?*, Brussels: Belgian Science Policy.

Pant, R. and Sammer, K. (2003) 'Labeling as an appropriate strategy and instrument for sustainable consumption', paper presented at the 6th Nordic Conference on Environmental Social Sciences, Turku, June.

Passet, R. (2001) *Eloge du mondialisme par un 'anti' présumé*, Paris: Fayard.

Pêche, T. and Padis, M.O. (2004) *Les multinationales du cœur. Les ONG, la politique et le marché*, Paris: Le Seuil.

Pedersen, L.H. (1999) *The Dynamics of Green Consumption*, ECEEE Summer Study Proceedings. Online. Available: http://www.eceee.org/library_links/proceedings/1999/pdf99/Panel3/3-23.pdf.

Pellemans, P. (1999) *Recherche qualitative en marketing: perspective psychoscopique*, Paris: De Boeck.

Persais, E. (1998) 'La relation entre l'organisation et son environnement: le cas de l'écologie', *Revue française du Marketing*, 167: 27–44.

Pessin, A. (2001) *L'imaginaire utopique aujourd'hui*, Paris: PUF.

Phillips, D.L. and Clancy, K.J. (1972) 'Some effects of "social desirability" in survey studies', *American Journal of Sociology*, 77: 921–940.

Pignarre, P. (2004) 'Faire entrer le médicament en politique : le rôle du public', in F.X. Verschave (ed.) *La santé mondiale entre racket et bien public*, Paris: Charles Leopold Mayer.

Pirotte, G. (2004), 'Commerce équitable cherche ligne de conduite', *Politique*, 36: 46–49.

Pogge, T. (1998) 'A global resources dividend', in D.A. Crocker and T. Linden (eds) *Ethics of Consumption. The Good life, Justice and Global Stewardship*, Lanham, MD: Rowman & Littlefield.

Poncelet, M. (ed.) (2005) *A Fair and Sustainable Trade, Between Market and Solidarity: Diagnosis and Prospects*, Brussels: Belgian Science Policy.

Poncelet, M., Defourny, J. and De Pelsmacker, P. (2003) *A Fair and Sustainable Trade: Diagnosis and Prospects*, unpublished intermediary report, Brussels: Belgian Science Policy.

Poortinga, W., Steg, L. and Vlek, C. (2004) 'Values, environmental concern and environmental behavior. A study into household energy use', *Environment and behavior*, 36: 70–93.

Princen, T. (1999) 'Consumption and environment: some conceptual issues', *Ecological Economics*, 31: 347–363.

Princen, T. (2002) 'Consumption and its externalities', in T. Princen, M. Maniates and K. Conca (eds) *Confronting Consumption*, Cambridge, MA: MIT Press.

Princen, T., Maniates, M. and Conca, K. (eds) (2002) *Confronting Consumption*, Cambridge, MA: MIT Press.

Raikes, P., Jensen, M.F. and Ponte, S. (2000) 'Global commodity chain analysis and the French *filière* approach: comparison and critique', *Economy and Society*, 29: 390–417.

Ramachandran, V. and Saihjee, A. (2001) *Flying with the Crane. Recapturing Ten Years' Journey of KMVS*, Bhuj: KMVS.

Raynolds, L.T. (2002) 'Consumer–producer links in Fair Trade coffee networks', *Sociologia Ruralis*, 42: 404–424.

Reisch, L. A. (2001) 'Time and wealth: the role of time and temporalities for sustainable patterns of consumption', *Time and Society*, 10: 367–385.

Reisch, L.A. and Ropke, I. (eds) (2004) *The Ecological Economics of Consumption*, Cheltenham: Edward Elgar.

Renard, M.-C. (1999) 'The interstices of globalisation: the example of fair coffee', *Sociologia Ruralis*, 39: 484–500.

Ritzer, R. (1998) *The McDonaldization Thesis: Explorations and Extensions*, London: Sage.

Roberts, J.A. (1995) 'Profiling levels of socially responsible consumer behavior: a cluster analytic approach and its implications for marketing', *Journal of Marketing*, autumn: 97–117.

Roberts, J.A. (1996) 'Will the real socially responsible consumer please step forward?', *Business Horizons*, 39: 79–83.

Rochefort, R. (1997) *Le consommateur entrepreneur. Les nouveaux modes de vie*, Paris: Odile Jacob.

Roozen, N. and Vanderhoff, F. (2002) *L'aventure du commerce équitable*, Paris: J.C. Lattès.

Røpke, I. (1999) 'The dynamics of willingness to consume', *Ecological Economics*, 28: 399–420.

Rotolo, T. (2000) 'A time to join, a time to quit: the influence of life cycle transitions on voluntary association membership', *Social Forces*, 78: 1133–1161.

Rousseau, J.J. (1754) *What is the Origin of Inequality among Men and is it Authorized by Natural Law?*, trans. G.D.H. Cole. Available: http://www.constitution.org/jjr/ineq.htm.

Ruano-Borbalan, J-C. (1998) *L'identité, l'individu, le groupe, la société*, Auxerre: Sciences Humaines Éditions.

Rubik, F. and Scholl, G. (2002) 'Integrated Product Policy (IPP) in Europe – a development model and some impressions', *Journal of Cleaner Production*, 10: 507–515.

Rumpala, Y. (1999) 'Le réajustement du rôle des populations dans la gestion des déchets ménagers', *Revue Française de Science Politique*, 49: 601–630.

Sabourin, M. and Lamarche, L. (2005) 'Environmental social psychology', in Psychology from *Encyclopedia of Life Support Systems (EOLSS)*, Oxford: EOLSS Publishers.

Sachs, W. (1992) *For Love of the Automobile*, Berkeley, CA: University of California Press.

Sachs, W. (1999a) 'Turning vision into reality: rethinking how sustainable business must operate in future', paper presented at FRDO-CFDD Symposium *'Politique de produits intégrée'*, Brussels, September.

Sachs, W. (1999b) *Planet Dialectics*, London: Zed Books.

Sachs, W., Loske, R. and Linz, M. (1998) *Greening the North. A Post-industrial Blueprint for Ecology and Equity*, London and New York: Zed Books.

Sahlins, M. (1974) *Stone Age Economics*, London: Tavistock Press.

Sandjuro, D. (2001) 'Quel contrôle pour le commerce équitable?', *Le Monde Diplomatique*, Supplément Economie solidaire, October.

Sanne, C. (2002) 'Willing consumers – or locked-in? Policies for a sustainable consumption', *Ecological Economics*, 42: 273–287.

Sardar, Z. (1998) *Postmodernism and the Other: The New Imperialism of Western Culture*, London: Pluto Press.

Sayer, A. (2001) 'For a critical cultural political economy', *Antipode*, 33: 687–708.

Scherer, I. (2004) 'Can an informative artefact induce sustainable behaviour in the French households? The answer of a cognitive transfer experience', unpublished Ph.D. thesis, Université de Liège, Belgium.

Schumacher, E.F. (1974) *Small is Beautiful*, London: Abacus.

Schutz, A. (1987) *Le chercheur et le quotidien. Phénoménologie des sciences sociales*, Paris: Méridiens Klincksieck.

Schwartz, N. and Strack, F. (1991) 'Evaluating one's life: a judgment model of subjective well-being', in F. Strack, M. Argyle and N. Schwartz (eds) *Subjective Well-being: An Interdisciplinary Perspective*, London: Pergamon Press.

Scott, J. C. (1997) 'The infrapolitics of subordinate groups', in M. Rahnema and V. Bawtree (eds) *The Post-development Reader*, London: Zed Books.

Segal, J. M. (1998) 'Consumer expenditures and the growth of needed-required income', in D.A. Crocker and T. Linden (eds) *Ethics of Consumption. The Good Life, Justice and Global Stewardship*, Lanham, MD: Rowman & Littlefield.

Semelin, J. (1989) *Sans armes face à Hitler – la résistance civile en Europe*, Paris: Payot.

Sharma, D. (2004) Governments change: but no respite from farmer's suicides. Available: http://www.mindfully.org/WTO/2004/India-Farmers-suicides3jun04.htm, (accessed January 2006).

Shaw, D. and Clarke, I. (1999) 'Belief formation in ethical consumer groups: an exploratory study', *Marketing Intelligence & Planning*, 17: 109–119.

Shove, E. (2001) *Notes on Comfort, Cleanliness and Convenience*, Lancaster: Lancaster University Press.

Shove, E. (2003) 'Changing human behaviour and lifestyle: a challenge for sustainable consumption?', London: PSI. Online. Available: http://www.psi.org.uk/ehb/docs/shove-changinghumanbehaviourandlifestyle-200308.pdf (accessed 3 March 2006).

Shove, E. (2005) 'Changing human behaviour and lifestyle: a challenge for sustainable consumption?', in I. Ropke and L. Reisch (eds) *Consumption – Perspectives from Ecological Economics*, Cheltenham: Edward Elgar.

Shove, E., Lutzenhiser, L., Guy, S., Hackett, B. and Wilhite, H. (1998) 'Energy and social systems', in S. Rayner and E. Malone (eds) *Human Choice & Climate Change. Volume 2. Resources and Technology*, Ohio: Battelle Press.

Shreck, A. (2002) 'Just bananas? Fair trade banana production in the Dominican Republic', *International Journal of Sociology of Agriculture and Food*, 10: 11–21.

Siebenhüner, B. (2000) 'Homo sustinens – towards a new conception of humans or the science of sustainability', *Ecological Economics*, 32: 15–25.

Sikula, A. Sr. and Costa, A.D. (1994) 'Are women more ethical than men?', *Journal of Business Ethics*, 13: 859–871.

Siméant, J. (2001) 'Entrer, rester en humanitaire: des fondateurs de MSF aux membres actuels des ONG médicales françaises', *Revue Française de Science Politique*, 51: 47–74.

Singh, S.N. and Cole, C.A. (1993) 'The effects of length, content, and repetition on television commercial effectiveness', *Journal of Marketing Research*, 30: 91–105.

Smith, G.A. (1993) 'The purpose of wealth: a historical perspective', in H.E. Daly and K.N. Townsend (eds) *Valuing the Earth*, Cambridge, MA: MIT Press.

SONECOM (2005) *Enquête d'opinion sur le commerce équitable*, Brussels: CTB.

Southerton, D., Chappells, H. and Van Vliet, B. (eds) (2004) *Sustainable Consumption: The Implications of Changing the Infrastructures of Provisions*, London: Edward Elgar.

Spaargaren, G. (1997) 'The ecological modernization of production and consumption. Essays in environmental sociology', unpublished thesis, Landbouwuniversiteit Wageningen.

Spaargaren, G. (1999) 'The ecological modernization of domestic consumption', Paper presented at Everyday Life and Sustainability Summer School, Lancaster University.

Spaargaren, G. (2000) 'Ecological modernization theory and domestic consumption', *Journal of Environmental Policy and Planning*, 2: 323–335.

Spangenberg, J.H. and Lorek, S. (2002) 'Environmentally sustainable household consumption: from aggregate environmental pressures to priority fields of action', *Ecological Economics*, 43: 130–131.

Spence, A. M. (1974) *Market Signalling: Information Transfer in Hiring and Related Process*, Cambridge, MA: Harvard University Press.

Spencer, H. (1851) *Social Statistics*, New York: Appleton.

Stagl, S. and O'Hara, S.U. (2001) 'Preferences, needs and sustainability', *International Journal of Sustainable Development*, 4: 4–21.

Stangherlin, G. (2004) 'Une approche multidimensionnelle et processuelle du militantisme. L'engagement pour l'autre lointain dans les ONG de coopération au développement en Belgique', unpublished thesis, Université de Liège, Belgium.

Stearns, P.N. (2001) *Consumerism in World History*, London and New York: Routledge.

Steg, L. and Buijs, A. (2004) *Psychologie & duurzame ontwikkeling, Vakreview*, Nijmegen: Werkgroep Disciplinaire verdieping Duurzame Ontwikkeling.

Stengers, I. (1993) *L'invention des sciences modernes*, Paris: La Découverte.

Stern, P.C. (1992) 'Psychological dimensions of global environmental change', *Annual Review of Psychology*, 43: 269–302.

Stern, P.C. (1997) 'Toward a working definition of consumption for environmental research and policy', in P.C. Stern, T. Dietz, V.W. Ruttan, R.H. Socolow and J.L. Sweeney (eds) *Environmentally Significant Consumption*, Washington, DC: National Academy Press.

Stern, P.C. (2000) 'Towards a coherent theory of environmentally significant behaviour', *Journal of Social Issues*, 56: 407–424.

Stewart, D.W. and Shamdasani, P.N. (1990) *Focus Groups: Theory and Practice*, London: Sage.

Stø, E. and Strandbakken, P. (2002) 'Advantages and limitations of ecolabels as consumer and environmental political instruments', Paper presented at IIASA workshop 'Lifecycle Approaches to Sustainable Consumption', Laxenburg, Austria, November.

Stø, E., Strandbakken, P., Throne-Holst, H. and Vittersø, G. (2004) 'Potentials and limitations of environmental information to individual consumers', Paper presented at the 9th European Roundtable on Sustainable Consumption and Production, Bilbao, Spain, May.

Szykman, L.R., Bloom, P.N., and Levy, A.S. (1997) 'A proposed model of the use of package claims and nutrition', *Journal of Public Policy and Marketing*, 16: 228–241.

Tallontire, A. (2001) *Fair Trade and Development, The guide to developing markets and agro-enterprises*, National Resources Institute (NRI) and World Bank.

Tallontire, A., Rentsendorj, E. and Blowfield, M. (2001) *Ethical Consumers and Ethical Trade: A Review of Current Literature*, Policy Series 12, London, Natural Resources Institute, University of Greenwich. Medway: University of Greenwich.

Tap, P. (1998) 'Marquer sa différence', in J-C. Ruano-Borbalan (ed.) *L'identité, l'individu, le groupe, la société*, Paris: Sciences Humaines.

Teisl, M.F., Roe, B., and Levy, A.S. (1999) 'Eco-certification: why it may not be a field of dreams', *American Journal of Agricultural Economics*, 81: 1066–1071.

The Roper Organization Inc (1990) *The Environment: Public Attitudes and Individual Behavior*, Storrs, CT: University of Connecticut.

Thorgersen, J. (2000) 'Psychological determinants of paying attention to eco-labels in purchase decisions: model development and multinational validation', *Journal of Consumer Policy*, 23: 285–313.

Thorgersen, J. and Olander, F. (2002) 'Human values and the emergence of a sustainable consumption pattern: a panel study', *Journal of Economic Psychology*, 23: 605–630.

Thuderoz, C., Mangematin, V. and Harrisson, D. (1999) *La confiance. Approches économiques et sociologiques*, Paris and Quebec: Gaëtan Morin.

Ton, P. (2002) *The International Market for Organic Cotton and Organic Textile*. Online. Available: http://www.pan-uk.org/Cotton/Markets.pdf (accessed March 2005).

Touraine, A. (1978) *La voix et le regard*, Paris: Seuil.

Tsalikis, J. and Ortiz-Buonafina, M. (1990) 'Ethical beliefs' differences of males and females', *Journal of Business Ethics*, 9: 509–517.

UNCED (United Nations Conference on Environment and Development) (1992) *Rio Declaration on Environment and Development*.

UNDP (United Nations Development Programme) (2003) *Human Development Report 2003*, New York: Oxford University Press.

United Nations Environment Program (2002) *Global Environment Outlook 3*, Stevenage: Earthprint.

United Nations Environment Program (2004) *Informal Meeting of an Advisory Task Force on the 10-year Framework of Programmes on Sustainable Consumption and Production*, Background Note.

Vaillancourt, J-G. (1996) 'Sociologie de l'environnement: de l'écologie humaine à l'écosociologie', in R. Tessier and J-G Vaillancourt (eds) *La recherche sociale en environnement. Nouveaux paradigmes*, Montréal: Les Presses de l'Université de Montréal.

Van Campenhoudt, L. (2001) Introduction à l'analyse des phénomènes sociaux, Paris: Dunod.

Van den Bergh, J.C.J.M. and Ferrer-i-Carbonell, A. (2000) 'Economic theories of sustainable consumption', in B. Heap and J. Kent (eds) *Towards Sustainable Consumption, A European Perspective*, London: The Royal Society.

Vandercammen, M. (2002) *Comportements du consommateur*, Brussels: CRIOC.

Veblen, T. (1899) *The Theory of Class Leisure*, Harmondsworth: Penguin Classics.

Vitell, S., Singhapakdi, J. and Thomas, J. (2001) 'Consumer ethics: an application and empirical testing of the Hunt–Vitell Theory of Ethics', *Journal of Consumer Marketing*, 18: 153–178.

Vitousek, P.M. and Mooney, H.A. (1997) 'Human domination of Earth's ecosystems', *Science*, 277: 494–499.

Viveret, P. (2002) 'Reconsidérer la richesse', report to the Secrétaire d'Etat à l'Economie solidaire Guy Hascoët, Paris.

Vivien, F-D. and Pivot, A. (1999) 'A propos de la méthode d'évaluation contingente', *Natures Sciences Sociétés*, 7: 33–64.

Vlasselaer, M. (1997) *Le pilotage d'entreprise. Des outils pour gérer la performance future*, Louvain-la-Neuve: Publi-Union Editions.

von Weiszäcker, E-U., Lovins, A.B. and Lovins, L.H. (1997) *Factor Four. Halving Resource Use. The New Report to the Club of Rome*, London: Earthscan.

Wachtel, P.L. (1983) *The Poverty of Affluence*, New York: Free Press.

Warlop, L., Vanden Abele, P. and Smeesters, D. (2001) *Between Green Thoughts and Green Deeds: the Relationship Between Environmental Concern and Source Separation Performance for Individual Consumers: the final report*, Brussels: Politique Scientifique.

Wasserman, D. (1998) 'Consumption, appropriation and stewardship', in D.A. Croker and T. Linden (eds) *Ethics of Consumption. The Good Life, Justice and Global Stewardship*, Lanham, MD: Rowman & Littlefield.

Weber, M. (1978) *Economy and Society, an Outline of Interpretative Sociology*, Berkeley, CA: University of California Press.

Wessels, C.R., Johnston, R.J. and Donath, H. (1999) 'Assessing consumer preferences for eco-labelled seafood: the influence of specie, certifier and household attributes', *American Journal of Agricultural Economics*, 81: 1084–1089.

Whatmore, S. and Thorne, L. (1997) 'Nourishing networks: alternate geographies of food', in D. Goodman and M. Watts (eds) *Globalizing Food, Agrarian Questions and Global Restructuring*, London and New York: Routledge.

Wilk, R. (2001) 'Consuming morality', *Journal of Consumer Culture*, 1: 245–260.

Wilk, R. (2002) 'Consumption, human needs, and global environmental change', *Global Environmental Change*, 12: 5–13.

Wilkinson, R. G. (1973) *Poverty and Progress*, New York and London: Praeger Press.

Williger, M. (1999) 'La méthode d'évaluation contingente: de l'observation à la construction des valeurs de préservation', *Natures Sciences Sociétés*, 4, 1, Paris.

Wilting, H.C., Biesiot, W. and Moll, H.C. (1999a) 'Analyzing potentials for reducing the energy requirement of households in The Netherlands', *Economic Systems Research*, 11: 233–243.

Wilting, H.C., Moll, H.C. and Nonhebel, S. (1999b) 'An integrative assessment of greenhouse gas reduction options', *IVEM Research Report*, 101, Groningen: University of Groningen, Centrum Voor Energie en Milieukunde IVEM.

WCED (World Commission on Environment and Development) (1987) *Our Common Future*, Oxford: Oxford University Press.

Worldshops (2005) Available: http://www.worldshops.org, accessed 2 February 2005.

Zaccaï, E. (2000) 'Ecological oriented consumption: a pluriactoral approach', *International Journal of Sustainable Development*, 3: 26–39.

Zaccaï, E. (2002) *Le développement durable. Dynamique et constitution d'un projet*, Bern and Brussels: PIE-Peter Lang.

Zaccaï, E. (2003) 'Changing unsustainable patterns of consumption and production', in E. Nierynck, A. Vanoverschelde, T. Bauler, E. Zaccaï, L. Hens and M. Pallemaerts (eds) *Making Globalization Sustainable*, Brussels: Brussels University Press.

Zaccaï, E. (2006) 'Assessing the role of consumers in sustainable product policies', *Environment, Development and Sustainability*. Available: http://www.springerlink.com, reference no. SSN: 1387-585X (Paper); 1573-2975 (online) pp. 1–17.

Zadek, S., Lingayah, S. and Forstater, M. (1998) *Social Labels: Tools for Ethical Trade*, London: New Economics Foundation.

Zelizer, V.A. (2005) 'Circuits within capitalism', in V. Nee and R. Swedberg (eds) *The Economic Sociology of Capitalism*, Princeton, NJ: Princeton University Press.

Index

Segal, J.M. 21, 22
segmentation 111–13
selective motivations 74
self-evaluation 96–8
self-reflexive attitude 94–107
sensitive consumer 233
sewers and sewing 221, 222
shareholding democracy 178–9
Shaw, D. 114
short producer–consumer chains 180–1
Shrujan 218
signalling theory 196
situations, and practices 64–6, 68–9
small producers 203–4
small retailers 23
small shareholders 179
social capital 23
social change 13, 186–200; changes in
    individual consumption depend on
    organisation of collective action
    192–8; individual action's meaning
    for a social group 188–92; towards
    more sustainable modes of
    consumption 186–8
social commitment *see* commitment
social congestion 50
social consumerism movement 193
social desirability effect 46
social dilemma 75
social dimension of sustainable
    development 76–7; social criteria
    237
social expectations 190–1
social justifications 154–5, 158–9
social label 119–23
social observation agencies 197
social performance management system
    196
social relations 163–4, 175–6
social standards 225–6
social vulnerability 174–6
sociology: consumers defined by
    practices and situations 64–6, 68–9;
    definition of consumption 35;
    sociological profiles of World Shops
    customers 131, 132, 133–4
socio-economic class 175–7
socio-technical network 216–30
socio-technology system 66–7

solidarity 78–80, 81, 147; and buying
    Fair Trade products 136–7, 139–40,
    181, 233–4; mobilising and
    interpretative frames 189–92, 194–5
solidarity relationships, typology of
    136–7, 138–9, 233–4
Solidar'Monde 202, 204–5, 209–10,
    212, 213
sorting of waste 91, 94–107
South–North relations 26–7, 128,
    130–1, 138, 140, 184, 217
Southern producers: impact of Fair
    Trade on 13, 216–30; *see also*
    large-scale Fair Trade
Spaargaren, G. 37
space 78–80, 81
spare time 175–6
speciality shops 117, 123
Spence, A.M. 196
Spencer, H. 25–6
stakeholders 187, 195–8
standards 3, 6; environmental 87;
    quality 220–2; social 225–6
statistics 173–4
status 49
Steg, L. 38
Stern, P.C. 35, 62, 74
store brands 122, 123, 125
strategies for changing consumption
    80–1, 89
structural strategies 38–9
structuration 65
subjective well-being (SWB) 44–7;
    criticisms and limitations of
    approach 46–7; and income 44–6
supermarkets 116–17, 123–4, 125, 211
sustainability 5, 30–1, 35, 53, 186;
    criteria of ecological and social
    sustainability 236–8; and
    overconsumption 24–5
sustainable decrease 156–8
sustainable development 30–1, 156–7
Sweden 36–7
system, relationship of individual with
    153
Szykman, L.R. 113

targeting 111–13
task-based motivation profiles 81–2